Routledge Revivals

U.S. Household Consumption, Income, and Demographic Changes

The impacts of the two variables of population and income growth on resources and the environment are transmitted through their effects on the demands for goods and services. To enrich our understanding of the impacts of population and income on consumer demand, Philip Musgrove, with the assistance of Adele Shapanka, undertook the research in this volume, which was first published in 1982. This book will be of interest to students of economics and environmental studies.

UK's Household Consumption,
Incomes and Demographic Change

U.S. Household Consumption, Income, and Demographic Changes

1975-2025

Philip Musgrove

With the Assistance of
Adele Shapnanka

RFF PRESS
RESOURCES FOR THE FUTURE

First published in 1982
by Resources for the Future, Inc.

This edition first published in 2015 by Routledge
2 Park Square, Milton Park, Abingdon, Oxon, OX14 4RN
and by Routledge
711 Third Avenue, New York, NY 10017

Routledge is an imprint of the Taylor & Francis Group, an informa business

© 1982 Resources for the Future, Inc.

Publisher's Note
The publisher has gone to great lengths to ensure the quality of this reprint but points out that some imperfections in the original copies may be apparent.

Disclaimer
The publisher has made every effort to trace copyright holders and welcomes correspondence from those they have been unable to contact.

A Library of Congress record exists under LC control number: 81086060

ISBN 13: 978-1-138-93464-1 (hbk)
ISBN 13: 978-1-315-67767-5 (ebk)
ISBN 13: 978-1-138-93467-2 (pbk)

U.S. Household Consumption, Income, and Demographic Changes
1975–2025

Philip Musgrove

with the assistance of

Adele Shapanka

RESOURCES FOR THE FUTURE / WASHINGTON, D.C.

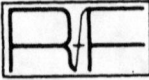

RESOURCES FOR THE FUTURE, INC.
1755 Massachusetts Avenue, N.W., Washington, D.C. 20036

Resources for the Future is a nonprofit organization for research and education in the development, conservation, and use of natural resources and the improvement of the quality of the environment. It was established in 1952 with the cooperation of the Ford Foundation. Grants for research are accepted from government and private sources only on the condition that RFF shall be solely responsible for the conduct of the research and free to make its results available to the public. Most of the work of Resources for the Future is carried out by its resident staff; part is supported by grants to universities and other nonprofit organizations. Unless otherwise stated, interpretations and conclusions in RFF publications are those of the authors; the organization takes responsibility for the selection of significant subjects for study, the competence of the researchers, and their freedom of inquiry.

Research Papers are studies and conference reports published by Resources for the Future from the authors' typescripts. The accuracy of the material is the responsibility of the authors and the material is not given the usual editorial review by RFF. The Research Paper series is intended to provide inexpensive and prompt distribution of research that is likely to have a shorter shelf life or to reach a smaller audience than RFF books.

Library of Congress Catalog Card Number 81-86060

ISBN 0-8018-2868-6

Copyright © 1982 by Resources for the Future, Inc.

Distributed by The Johns Hopkins University Press,
  Baltimore, Maryland 21218

Manufactured in the United States of America

Published January 1982

CONTENTS

FOREWORD

This book originated as part of a larger study undertaken at RFF under the direction of Ronald G. Ridker to examine the implications of alternative developments on resources and the environment over the half century, 1975-2025. The principal findings of that larger RFF study were published in To Choose A Future: Resource and Environmental Consequences of Alternative Growth Paths by Ronald G. Ridker and William D. Watson (Baltimore, Johns Hopkins Press for Resources for the Future, 1980).

Population and income growth are the principal independent variables examined in the Ridker-Watson book. The impacts of the two variables on resources and the environment are transmitted through their effects on the demands for goods and services. To enrich our understanding of the impacts of population and income on consumer demand, Philip Musgrove, with the assistance of Adele Shapanka, undertook the research reported in this volume.

The assumptions and framework of the demand analysis of this study were adapted to complement the broader study by Ridker and Watson. Both the sectoral breakdown and the population growth assumptions are identical to those used in To Choose A Future. Important differences between the studies involve their treatment of income growth and population composition. In order to focus on population effects, Musgrove examines one set of scenarios assuming no growth in labor productivity, a set not used by Ridker and Watson, and another set assuming annual productivity growth of slightly less than 2 percent, which is used in the earlier book. The composition of population, an item of major interest in the Musgrove study, is not treated by Ridker and Watson.

A principal contribution of the Musgrove study is its use of cross-sectional data to derive consumption functions by categories of households. This study goes well beyond previous efforts in its use of the

1972-1973 Consumer Expenditure Survey (CES) of the Bureaus of Census and
Labor Statistics to analyze consumption. Detailed expenditure functions
sensitive to the size of a household, the age of the household head, and
the adult/child composition of a household are derived, and the number of
each type of household as described by these variables, is projected. As
a result, the analysis partitions projected changes in expenditure into
three contributing components--increases in population, growth of income
per person, and changes in the composition of population by age and by
grouping into households.

The cross-sectional data, however, do not fully cover expenditures on
education and health care, and in some cases the results obtained with the
cross-sectional data were discarded as unsatisfactory. Consequently, a
combination of approaches was used to estimate consumption, and in the
final results less than half of the expenditures are estimated from the CES
data. Other results are derived from the use of Ridker-Watson aggregate,
time series equations or are developed separately from demand models for
education, health care, and housing construction.

Since the analysis is limited to a comparison of scenarios which differ significantly only with respect to population and income growth,
factors such as tastes, prices, and technical progress are invariant among
scenarios. Over the time horizons considered in the study, these factors
can be expected to have major impacts on consumer demand. Thus, while the
results provide interesting insights into the role of demographic variables
on consumer demand, it should be emphasized that these estimates, particularly those based on the CES data, should not be considered as forecasts
but rather as sources of insights as to the long-run expenditure implications resulting from changes in a limited number of key variables such as
income and population size and composition, including household characteristics.

Because of their bulk, the detailed consumption functions are not
reproduced in this volume. The interested reader may obtain the tables
providing the consumption equation coefficients and standard errors for
each sector from the Publications Office of Resources for the Future.

Despite the qualifications of the results as forecasts, the study does
make a major contribution to the methodology and analysis of consumer
expenditures and projections of consumption patterns. Transformation of

the Consumer Expenditure Survey data into the household categories used in the analysis and into a form compatible with the input-output model used in the Ridker-Watson study represents a major and innovative research task. Furthermore, the results suggest some interesting conclusions as to the role of demographic composition on expenditure patterns which can be tested by further research. For these reasons, RFF is pleased to disseminate the research by publishing this volume.

Philip Musgrove directed this study and wrote Chapters 1, 2 and 6, which summarize the study and present the private consumption results. Adele Shapanka, who worked with Ridker and Watson on the larger project, contributed parts of Chapter 2, developed the projections of public consumption and construction in Chapters 3, 4, and 5, and wrote first drafts of those chapters.

Marcia Mason provided expert help with data handling and computation. She was responsible for all the computer work except the preparation of final tables, which were created by Paul Morawski. Carolyn Cummings-Saxton assisted with the detailed work of preparing the consumer expenditure data, including numerous adjustments and reconciliations with other sources. John Mankin typed the several versions of the manuscript. Ronald Ridker and William Watson provided general advice and guidance. Irving Hoch read the initial manuscript and was generous and thorough in his suggestions for revision. The final form of the study owes much to his perceptive criticism. Thomas Lareau and Jack Alterman commented extensively on a later version of the manuscript. Ruth Haas also offered helpful comments. In addition to the Center for Population Research, the Department of Energy helped finance this project by providing free computer facilities.

Kenneth D. Frederick, Director
Renewable Resources Division
Resources for the Future

November, 1981

Chapter 1

INTRODUCTION AND SUMMARY

The fundamental question addressed by this study is, how might different courses of population growth over the next half century affect the
level and structure of final demand in the United States economy? The
bulk of final demand is for personal consumption, that is, expenditures
made directly by households for their own use. A priori, any element of
this expenditure may be sensitive to demographic growth, since different
kinds of households may have different spending patterns. Public consumption, which constitutes the rest of consumption demand, includes some elements which may be expected to be very demographically sensitive, such as
education, and others which probably do not vary with the composition of
the population and may even be fairly insensitive to its size, such as
defense. Most of investment, the third major component of domestic product, is a derived demand, undertaken in response to household or public
demands for goods and services, but it includes one component--housing
construction--which is bought directly by households and may be expected
to vary significantly with demographic factors. Two other components of
construction--for schools and for hospitals--are derived directly from
current and projected expenditures on education and on health care, respectively, and so may also be demographically sensitive.

Since our emphasis is on demographic effects on demand, we are concerned with the demands for goods and services consumed by individuals
(whether or not they buy the goods and services), and not with the consumption of indivisible public goods. Nearly all of private consumption
in this sense is accounted for by personal consumption expenditure; the
remainder consists of publicly provided education and health care. These
last two components are traditionally classified as government expendi-

tures, but we include them in our enlarged definition of consumption. The construction components of interest--housing, schools and hospitals-- also mix public and private investments.

In this study, we examine the consequences of population alternatives on private consumption and these three types of construction only, without deriving from them the level and composition of other types of investment (buildings, machinery and transportation infrastructure) that would be required. We also do not attempt to study saving, international trade, or the effects of different demands on resource use or the emission of wastes. We also limit the analysis to comparisons among scenarios, or hypothetical futures, which differ significantly only in the assumptions made about population and income growth. All other factors affecting future demand-- tastes, prices, and technical progress--are assumed the same in all sce- narios, so that while they are extremely important, they contribute little or nothing to differences between one possible future and another. In particular, we assume that prices are independent of demand levels, so that supply constraints are not considered. This is unquestionably the most serious limitation of the study: it means that we are trying to iso- late certain effects of interest rather than project what the future will actually be.

Background to This Study

The general question of how population growth, and particularly changes in growth, affects final demand, has been investigated in a number of papers, many of which consider questions excluded from our analysis, such as investment and saving. Serow and Espenshade (1977) review this literature, and Espenshade (1977) has estimated demand functions for eleven expenditure categories, taking account of the population age dis- tribution (in three size classes) and the average household size. No household-level data were used.

The present analysis builds on the INFORUM model of the U.S. economy developed by Almon and associates (1974). That model divides personal consumption expenditures into 133 sectors related through an input-output table; public or governmental consumption comprises another fourteen sectors. Demand equations were originally fitted for each sector using aggregate time-series data, expenditure in a sector depending on total population, per capita disposable income and (for many sectors) a time trend to capture technological or taste changes. Disaggregated data entered into these estimates and projections in only two ways: first, income elasticities for some sectors were estimated from the 1960-61 Consumer Expenditure Survey of households; and second, demand in some sectors (particularly education) depended on population in specific age groups rather than on total population. This approach can in principle capture the effects of population scale and income on demand, but--with the exceptions noted--cannot say anything about compositional effects. Even when the age composition of the population is taken into account, no assumptions are made about the distribution of people, or of income, among households. To the extent that composition and distribution matter, this aggregate approach must erroneously attribute their effects either to scale or to per capita income changes, and it may wrongly estimate the total effect. In an extension of this work, Herzog (n.d.) used disaggregated data from the 1960-61 Consumer Expenditure Survey to make the income coefficients of the sectoral equations "demographic specific." The basic equation is still estimated from time-series data in this approach, but the income effect itself is related to population composition.

The INFORUM model was extended in a different way by Ridker and Watson (1980), who re-estimated a number of the sectoral demand equations or modified them by changing the assumptions about time trends. The revised set of demand functions was combined with three alternative projections of population growth, from Census Bureau series D, E and F, and with alternative assumptions concerning average labor productivity. Whereas Almon's group initially made projections only to the year 1985, Ridker and Watson carried their projections out to 2025, extending the Census Bureau's projections where necessary. The model was then applied to estimate the resource-use and environmental impacts of the three projected population scenarios. The underlying demand relations, although sometimes modified to take account of changes in prices, technology, tastes or other factors, are still based on aggregate, time-series data on consumption and income. These demand functions do not appear, sector by sector, in Ridker and

Watson's published study; we include the detailed results, for some sectors, here for the first time.

In 1972-73 the Census Bureau and the Bureau of Labor Statistics repeated a Consumer Expenditure Survey, providing detailed data on household spending patterns throughout the United States for the first time in more than a decade. These data make it possible to estimate demand in the sectors bought by households, at the level of individual households or spending units, taking account of their composition and of the distribution of income among them. The object of the present study, then, is to re-estimate the demand equations using these data with the greatest disaggregation possible, and to use them to project demand as a function of the projected structures of income and population. We continue to use the sectoral disaggregation employed by Ridker and Watson, which is shown in table 1-1. We also use their assumptions concerning population growth, and we assume alternatively that labor productivity does not grow at all (an assumption not considered by Ridker and Watson) or that it grows at a constant rate of just under two percent per year. Since our emphasis is on population rather than income effects, it did not seem worthwhile to experiment with other assumptions about income growth, and in particular with a more rapid growth of productivity. The assumptions of Ridker and Watson, together with our no-growth assumption, lead to a set of six projections (Census Bureau population series D, E and F, each with and without productivity growth, over the period 1975-2025). This framework is retained despite the fact that the Census Bureau has since revised downward slightly its population projections (new series I, II and III), for two reasons. One is that we are saved the trouble of creating new income projections to match the new population projections, which does not seem justified in view of the very small changes in the latter. The other is that we can compare our results directly with those obtained by Ridker and Watson, and thus see how much difference it makes to introduce demographic and distributional detail into the analysis. This provides a way to evaluate the implicit assumption made in all aggregative models, which is that distribution, either of people or of income, does not matter.

5

Table 1-1. Summary of Type of Equation Used for Sector Projections

Four types of projecting equations are distinguished:

(1) Time-series equation retained (from Ridker and Watson), not sensitive
   to population composition

(2) Household equation estimated from the CES, without subsequent adjustment

(3) Household CES equation, with adjustment overall or to individual coefficients. For types (2) and (3), D means the CES Diary data were used,
   and I indicates the use of the interview data

(4) New cross-section or time-series equation estimated from other sources
   than the CES (chiefly time-series data), sensitive to population composition

I. Personal Consumption Expenditure (PCE) Sectors

No.	Sector Description	Equation type:	(1)	(2)	(3)	(4)
1	Dairy farm products		x			
2	Poultry & eggs				xD	
3	Meat & animals		x			
7	Fruits, vegetables			xD		
8	Forestry & fishery products		x			
0	Agricultural, forestry & fishery		x			
4	Coal mining		x			
6	Stone & clay mining		x			
1	Ammunition		x			
2	Other ordnance		x			
3	Meat products				xD	
4	Dairy products				xD	
5	Canned & frozen foods			xD		
6	Grain mill products			xD		
7	Bakery products			xD		
8	Sugar				xD	
9	Confectionary products				xD	
0	Alcoholic beverages			xD		
1	Soft drinks & flavorings				xD	
2	Fats & oils				xD	
3	Miscellaneous food			xD		
4	Tobacco products				xD	
5	Broad & narrow fabrics				xI	
6	Floor coverings		x			
7	Miscellaneous textiles			xD+I		
8	Knitting			xI		
9	Apparel			xI		
0	Household textiles			xD+I		
1	Lumber & wood products		x			
3	Millwork			xI		
5	Household furniture					xI

No.	Sector Description	Equation type: (1)	(2)	(3)	(4)
46	Other furniture	x			
48	Paper & paperboard products	x			
49	Paper products, n.e.c.		xD+1		
51	Paperboard containers	x			
52	Newspapers			xI	
53	Periodicals		xI		
54	Books		xI		
55	Industrial chemicals	x			
56	Business forms		xD		
57	Commercial printing		xD		
58	Miscellaneous printing		xD		
59	Fertilizers		xI		
60	Pesticides	x			
61	Miscelleneous chemicals			xD	
62	Plastic materials	x			
66	Drugs				x
67	Cleaning & toilet preparations		xD		
68	Paints	x			
69	Gasoline	x			
70	Heating oil	x			
72	Tires & tubes		xI		
73	Rubber products		xD+I		
74	Miscellaneous plastics			xD+I	
76	Leather footwear		xI		
77	Other leather products		xI		
78	Glass	x			
80	Pottery			xI	
81	Cement, concrete & gypsum		xI		
82	Other stone & clay products		xD		
83	Steel	x			
87	Aluminum		xD		
90	Non-ferrous wire drawing	x			
93	Metal barrels, drums & rails	x			
94	Plumbing & heating equipment	x			
95	Structural metal products	x			
96	Screw machine products	x			
97	Metal stampings		xI		
98	Cuttery, hand tools & hardware	x			
99	Miscellaneous fabricated wire	x			
101	Other fabricated metal	x			
102	Engines & turbines	x			
103	Farm machinery		xI		
106	Machine tools	x			
108	Other metal working machinery	x			
109	Special industrial machinery	x			
115	Other office machinery	x			
116	Service industry machinery		xI		
117	Machine shop products	x			
119	Transformers & switchgear	x			
120	Motors & generators	x			

No.	Sector Description	Equation type: (1)	(2)	(3)	(4)
122	Welding apparatus		xI		
123	Household appliances		xI		
124	Electric lighting & wiring		xD+I		
125	Radio & television receiving			xI	
126	Phonograph records		xI		
127	Communication equipment	x			
128	Electronic components		xI		
129	Batteries	x			
130	Engine electrical equipment		xI		
131	X-ray equipment		xI		
133	Motor vehicles & parts			xI	
134	Aircraft	x			
137	Ship & boat building & repair	x			
139	Cycles & parts		xI		
140	Trailer coaches			xI	x
142	Mechanical measuring devices		xI		
143	Optical & ophthalmic goods				x
144	Medical & surgical instruments				x
145	Photographic equipment		xI		
146	Watches & clocks			xI	
147	Jewelry & silverware			xI	
148	Toys, sporting goods, musical instruments			xI	
149	Office supplies	x			
150	Miscellaneous manufacturing	x			
151	Railroads		xI		
152	Buses			xI	
153	Trucking		xI		
154	Water transportation			xI	
155	Airlines		xI		
157	Travel agents		xI		
158	Telephone & telegraph			xI	
160	Electric utilities	x			
161	Natural gas	x			
162	Water & sewer services		xI		
163	Wholesale trade } Estimated from trade margins				
164	Retail trade } of all PCE sectors				
165	Credit agencies & brokers		xI		
166	Insurance & brokers' agents		xI		
167	Owner-occupied dwellings	x			
168	Real estate			xI	
169	Hotels & lodging places			xI	
170	Personal & repair services		xD+I		
171	Business services		xI		
172	Advertising	x			
173	Auto repair	x			
174	Motion pictures & amusements		xI		
175	Medical services				x
176	Private schools		xI		x
177	Post office		xD		

No.	Sector Description	Equation type:	(1)	(2)	(3)	(4)
178	Federal government enterprises		x			
180	State & local electric utilities		x			
181	Directly allocated imports		x			

Total no. of PCE sectors is 133: ⎫7 sectors es-   51 51 25 6*
Estimated from interview data: 54 ⎬timated from
Estimated from diary data: 29   ⎭both sources

*Includes two sectors (140 and 176) for which a component is estimated from the CES, and the remainder from other sources. Shares of PCE for these sectors are included in column (4). Sectors 163 and 164 (trade margins) are not counted in any of the four equation categories, and account for 22.8 percent of base-year (1972) total PCE.

Share in base-year (1972) total PCE (percent)   18.48 25.18 22.32 11.24

II. Non-PCE Sectors Included in Total Expenditure

   Consumption

188	Health, welfare and sanitation (Health component only)			x
189	Public education			x

   Construction

18	New construction (Residential housing, schools and hospitals only)			x

In basing our analysis so far as possible on individual households, we are, of course, trading one set of problems for another, buying detail at the cost of possible biases and gaps. One immediate problem is that households do not account for all of spending on personal consumption, and, of course, they do not account for the rest of final demand. A hybrid analysis is necessary, in which some sectors will have to be estimated and projected from aggregate data, either because households do not buy for their own consumption or because the household information is inadequate. Some of these estimates are taken from Ridker and Watson, and some are newly estimated from other sources. A second difficulty is that cross-section estimates may be biased because they do not incorporate

time trends, representing as they do only one point in time, and the data
even at that point may contain transitory or atypical components of spending. Some link to aggregate estimates is therefore necessary to ensure
overall consistency. Finally, if the structures of population and income
determine the demand structure in the future, they must themselves be
projected, which is more difficult than projecting total population and
income. Assumptions must be made about behavior not only at the level of
spending but also at the level of household formation and the inter-
household distribution of income, and such assumptions leave much room
for error or doubt. Nonetheless, these assumptions are of interest in
their own right, and offer a way to see just how different population
scenarios might differ in detail. The rate of population growth cannot
change without also changing the age structure of the population, its
distribution into households, and the level and distribution of income.
It seems better to try to model these indirect consequences and to see
their effects on final demand than to suppress them, especially since
aggregate (non-distributional) projections are available to which to compare the results.

Organization of This Volume

As described earlier, the 1972-73 Consumer Expenditure Survey (CES)
provides the data from which demand is projected for the majority of consumption sectors. Since the projections are in constant 1971 producer
prices, while the data refer to consumer prices over a two-year period, a
great deal of adjustment is necessary before estimating demand functions.
A further complication arises because the CES is actually two independent
samples, one set of data referring to food and other frequent expenditures
with a weekly period of reference and the other to less frequent non-food
spending with an annual reference period. These two sources are linked in
the estimating equations via total food expenditures at home and away from
home, since these totals are reported in both surveys.

Chapter 2 describes the preparation of the CES data and the choice of
functional form for the demand equation, with the exact specification of
the four independent variables used: income, household size, household

composition and age of head. This chapter also describes how we projected the distributions of household size, composition and income. Since the estimation of the demand equations is itself a large task whose results may be of interest quite apart from their use for projection, the coefficients and standard errors of the equations are available on request. Chapter 2 summarizes the sizes and statistical significance of all the income and demographic effects, even for those equations which were subsequently modified for projection or replaced altogether. The reader interested only in the projection results may omit the latter parts of this chapter and proceed to chapter 6.

Several major spending categories which lie wholly or partly outside of personal consumption are discussed in separate chapters. Chapter 3 is devoted to educational spending, both for current operation (part of which is private and included in PCE, while the bulk is publicly provided) and for capital spending (school construction). Health care is treated in chapter 4: although current expenditures are ultimately all part of PCE, a substantial fraction are provided publicly and pass through government nondefense spending. Much of the private expenditure on health, in addition, is paid for by insurance, and so while it is bought for individuals it is not paid directly by them and does not appear in the CES data. Hospital construction is also projected in chapter 4. Finally, chapter 5 is devoted to housing construction, the one form of construction bought directly by households. The total spending of interest in this study is the sum of PCE plus current public educational and health spending plus the three construction sectors.

For educational spending, essentially the same procedure is followed for each of six levels of schooling, each level being associated with one and only one age group in the population (except that persons aged 18-34 may attend either colleges--undergraduate or graduate--or technical schools). First, an enrollment rate is projected, as a function of time or of GNP, using historical data. This allows economic conditions to influence school attendance, at least for those levels where saturation has not been reached. (Economic effects are strongest for college education, since it competes with labor force participation.) Enrollment follows from these rates and projected population in the appropriate age

group(s). Both elements of current expenditure--salary or personnel, and
nonsalary costs--are projected on a per-pupil basis as functions of GNP,
adjusted for full-time equivalence, and are then partitioned between public and private institutions. (No cost difference is assumed between the
types of school, at most levels, but private post-secondary schooling is
costlier.) Nonsalary spending is projected directly; salaries are found
by projecting personnel per pupil and applying a wage rate which grows at
the same rate as overall productivity, or--for comparison--is held constant. As with health care, construction expenditures are projected
separately, allowing for changes in the school age population and assuming
replacement rates for buildings. The data and the estimating equations
differ by school level within this general scheme. All the data are time-
series, so that the income, size and age composition of individual households do not enter the projections.

For health expenditures, cross-section (household survey) data are
used to determine age-specific rates of use of different kinds of medical
care, and time-series data are used to relate total health spending to
GNP. The elasticity derived in this way is assumed to shift downward
through time, and it is adjusted for demographic change in the past. Expenditures are then projected by combining the use rates, the projected
numbers of people in different age groups, and projected GNP. The projections of household composition enter this model only to adjust some use
rates for differences among household sizes. Projected health expenditure, in total and by sector, depends on income growth and on the population age structure, but only very little on household composition. The
arrangements for paying for medical care are assumed in all projections
to be the same as in 1975; the effects of existing public programs are
taken into account, but no assumptions are made concerning future changes
in public policy, such as the introduction of national health insurance.
Finally, expenditure for hospital construction is derived from spending
on (current) hospital care, using time-series data.

For housing construction, the projections depend on three elements.
The first is a matrix of housing choice, or types of housing (separate
house, apartment, mobile home) occupied by households classified according to size and age of head. This matrix is assumed constant, so that

while the actual housing mixture changes through time in response to
demographic change, any given type of household always has the same probability distribution over the three different types of dwelling. No account is taken of income or of factors such as urban/suburban/rural shifts
which might change these probabilities, which is admittedly a serious
limitation. However, no account is taken of such locational shifts for
any other sector, either. The second element is a set of equations relating house value (and hence cost of construction) to household income, given the choice of type. Thus, income growth will raise the value of the
housing stock even in the absence of demographic change. These first two
elements are both estimated from the CES data. The third element then is
exogenous estimates of the rates of abandonment and replacement of housing, leading from the stock to the amount of new construction necessary.
These estimates are based entirely on time-series relations.

Finally, chapter 6 presents the projection results, including both
longitudinal comparisons and comparisons among the three population scenarios, using the four years 1975, 1985, 2000 and 2025. Since the projections for different sectors are tied together via total projected PCE, we
present together, projections for all the component consumption sectors;
elements of education, medical care and housing are repeated from chapters
3 through 5 and combined with the sectors discussed in chapter 2. (The
reader interested only in the PCE sectors may proceed directly from chapter 2 to chapter 6.) As table 1-1 indicates, sector projections can be
of four different types, depending on the data source(s) used and on whether or not the demand equation derived in the cross-section is modified
for projection through time. At the end of this chapter we present a
brief summary of the findings reported in chapter 6, distinguishing among
the three principal effects on demand, due respectively to population
scale, composition, and income. We also attempt to see how Ridker and
Watson's time-series projections based on the same population and income
scenarios are changed by our introduction of demographic detail.

The Framework for Projection

The projection sequence is discussed in detail in chapter 2. Here
we present a brief summary of the principal elements. The three population scenarios to be compared correspond to Census Bureau series D, E and
F, extended to the year 2025. Total estimated population in that year is
367.5 million under scenario D, 303.8 million under E and 264.9 million
under F. The "given" information for each year in the projection interval
includes: the total population and its distribution over several age
brackets; the total number of households; the level of per capita disposable income; and the level of total personal consumption expenditure
(PCE). (The last variable is important because it is used to align our
projections with those made by Ridker and Watson, although our total expenditure concept does not correspond simply to PCE.) The remaining information necessary for the expenditure projections--the distribution of
people and of incomes among households--is itself projected, as described
in chapter 2. Results are given in the study for only the four benchmark years--1975, 1985, 2000, and 2025--although projections were made
year by year from 1972 to 1985, and at five-year intervals from 1985 to
2025.

The greatest difference among the three scenarios is in the assumed
rate of population growth, which is 0.64 percent faster per year, or more
than twice as fast, with scenario D than with F. Growth of income per
capita is inversely related to population growth, because of different
labor force participation rates, but the difference among the three scenarios is much smaller. At the end of the projection period, income per
capita in scenario F is 7.6 percent higher than in E, while in D it is
6.9 percent lower. In the year 2000, both differences are 6.0 percent.
Total PCE grows most rapidly under scenario D, the greater population
size outweighing the effect of a slightly slower growth of income.

For those sectors where demand is estimated from the Consumer Expenditure Survey, the projections of demand are made by combining two types
of estimates:

Demand functions are estimated for particular household types, as
functions of income and demographic characteristics. Projection for a

given type of household is then just a matter of substituting estimated income in the future year. Some of the projections for these sectors lead to expenditure estimates which seem unreasonably high or low; in those cases the demand equation is modified by suppressing or changing one or more coefficients, or by introducing a time trend. The basis for these modifications is sometimes taken from Ridker and Watson's estimates and sometimes from other sources: the demographic differences in the projections, both between scenarios and through time, remain. The modifications are described in chapter 6.

Distributional estimates are made first for households, and then for income distributed among households. The first gives the size distribution of households, the distribution of adults and children among them, and the age distribution of household heads. The second assigns an income to each kind of household as a function of mean income per capita and the household distribution. These estimates, of course, respect the given totals of households, adults, children, age groups and income.

The rules for projecting these two structures are identical across scenarios, but the results are not: the income distribution for one scenario, for example, is different from that for another in relative, not just absolute frequencies.

For sectors not based on the CES, we use either the aggregate equations of Ridker and Watson, or new demand equations based on other data. These may depend on income and total population alone, or may use some of the demographic detail such as the population age distribution or the household size distribution; none of them, however, depend on the distribution of income among households.

Finally, the projections of personal consumption spending by sector are summed to total PCE, which is then adjusted to match the total obtained for aggregate estimation. This adjustment is made by modifying projected income per capita so as to push the sectoral demands up or down until the totals are equal to within one percent. This procedure makes "income" an instrumental variable, decoupling it from the projected value based on productivity growth. This method creates no difficulties for projection because we do not try to model saving, and therefore can allow any sort of difference between total consumption and total income. The

non-PCE sectors studied are not involved in this adjustment process. The
same procedure is followed for both the income growth and no-growth scenarios. The entire sequence by which the sectoral projections are made
is shown diagrammatically in Figure 1-1. The interpretation in terms of
several effects on demand is discussed in the next section.

Factors Affecting Demand

We can classify the various determinants of final demand into three
fundamental factors: scale, composition and income. Scale refers to the
size of the population, so its effects simply shift all expenditures up or
down in the same proportion as population: by definition, they incorporate no changes in demand per person, and thus contribute nothing to changes in the structure of demand. Scale effects can be removed, for analysis, by expressing expenditures either in per capita terms or as shares
of total demand.

Composition refers to the age distribution of the population and the
way it is organized into households. The effects that go under this head-
ing in our analysis are the size distribution of households, their relative numbers of adults and children, the age distribution of household
heads and the distribution of income among households. (The last of these
is itself a function of the size and age distribution of households.)
Composition could be defined in many other, roughly equivalent, ways, but
this classification appears to capture the chief effects of interest.
Income and family size appear to be the chief determinants of spending in
most consumption categories. We estimate demand separately for each of
several household sizes, so as to take account of both factors without
having to assume any particular relation between them.

Many kinds of expenditure are age-sensitive, but there appear to be
two conceptually distinct age effects at work. One refers to the ages of
particular individuals, irrespective of their assignment to households:
this effect is most important for spending on education (concentrated in
the "school years") and on health care (concentrated both among the very
young, including the expense associated with birth, and among the elderly). The "needs" of these groups, defined by age, are largely indepen-

Figure 1-1. Summary Diagram of Projection Model, All Sectors

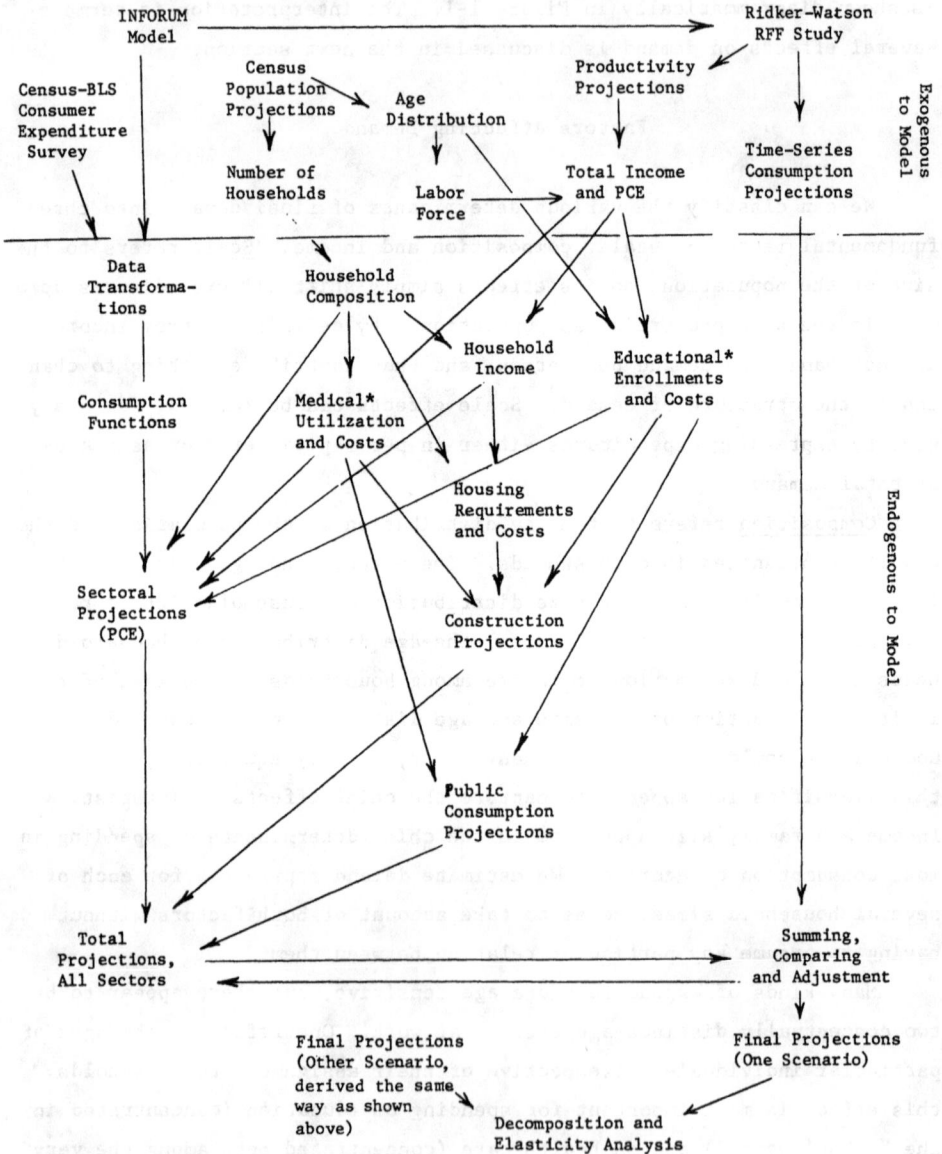

*Other exogenous data also used.

16

dent of household composition, and this independence is strongly reinforced by the fact that most of education and much of health care is publicly provided. The second kind of age effect refers to the "age" of the
entire household, and is usually represented by the age of the head. This
is, strictly speaking, a life-cycle effect, and it is likely to be most
important for goods and services used by the whole household rather than
bought for the consumption of particular members. Durable goods, including housing and vehicles, tend to show the strongest life-cycle effects.

We model the first kind of age effect, for most spending categories,
by distinguishing simply between adults and children, in households, with
the distinction at age 17/18. For education and health, however (chapters 3 and 4), we ignore households and consider the age composition of
the population, using finer age classes. For the second, or life-cycle
age effect, we consider only the age of the household head, and this variable is used to project demand in all sectors _except_ health and education.

Finally, _income_ is defined as income per capita (so as to separate
the scale effect), and can be expressed as the product of the share of the
population in the labor force and the average income generated by each
working person. (The income of the institutional population is not considered.) Because the level of income depends partly on labor force participation, it varies somewhat with the age composition of the population,
even if labor productivity is independent of demographic factors. Strictly speaking, therefore, the "income effects" of interest are effects of
productivity. On this definition, there are no "income differences" between two population growth paths, in a given year, if the same assumptions are made about productivity. Income differences arise only if different assumptions are made, or because productivity increases through
time.

These three factors can be combined in two different ways for analysis. One is to separate only the scale effects, so that the composition
of the population and the level of income together determine a _per capita_
effect, or equivalently, a _structural_ effect on demand. The other grouping combines the scale and composition factors into a total _demographic_
effect, leaving separate only the income effect:

$$\text{Total} \begin{array}{l} \nearrow \text{Scale (population)} \longrightarrow \text{Demographic} \\ \searrow \text{Per capita (structural)} \nearrow \text{Composition} \searrow \text{Income (per capita)} \end{array}$$

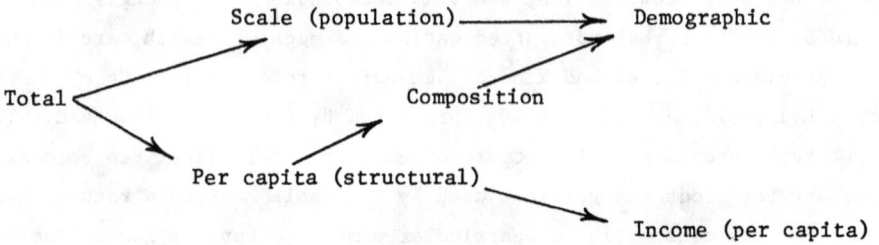

The three components of growth through time can be calculated as follows. Let N_0 and N_1 be population in a base year and a subsequent year, respectively; X_0 be base-year expenditure in a sector or category; X_1 be expenditure in the later year, assuming no income growth; and X_{1Y} be expenditure in the later year with income growth. The corresponding per capita levels are $\overline{X}_0 = X_0/N_0$, $\overline{X}_1 = X_1/N_1$, $\overline{X}_{1Y} = X_{1Y}/N_1$. Total growth of expenditure is $X_{1Y} - X_0$, which can be separated into three fractional components:

$$\frac{X_0 (N_1/N_0 - 1)}{X_{1Y} - X_0} \qquad \text{scale effect (SCA)}$$

$$\frac{X_1 - X_0 (N_1/N_0)}{X_{1Y} - X_0} \qquad \text{composition effect (COM)}$$

$$\frac{X_{1Y} - X_1}{X_{1v} - X_0} \qquad \text{income effect (INC)}$$

where $SCA + COM + INC = 1.0$. The sum of the first two terms gives the total demographic effect, the share in total growth of $X_1 - X_0$; the sum of the last two gives the total per capita or structural effect, the share in growth of $X_{1Y} - X_0 (N_1/N_0)$.

Dividing this last term by N_1 yields the total growth in per capita spending, $\overline{X}_{1Y} - \overline{X}_0$. Its two components (there is by definition no scale effect) are then:

$$\frac{\overline{X}_1 - \overline{X}_0}{X_{1Y} - X_0} \qquad \text{composition effect,} \quad \text{com} = \frac{\text{COM}}{1 - \text{SCA}}$$

$$\frac{\overline{X}_{1Y} - \overline{X}_1}{\overline{X}_{1Y} - \overline{X}_0} \qquad \text{income effect,} \quad \text{inc} = \frac{\text{INC}}{1 - \text{SCA}}$$

where the numerators are $(1/N_1)$ times the corresponding expressions for total growth.

Identifying these decompositions with the three effects of scale, composition and income is strictly valid only if no other factor is allowed to affect the demand function which generates X_1 and X_{1Y}. Any other such factor will be attributed to the two structural effects, depending on how it affects the subsequent expenditure levels, with and without income growth. In particular, changes in tastes or habits, or in relative prices, which are superimposed on the expenditure derived from scale, composition and income will alter the balance among the different effects. It should also be noted that while a negative scale effect is possible only if population shrinks or if projected expenditure is _less_ than that in the base year, and a negative income effect is possible only for an inferior good, there is no restriction on the sign of the composition effect.

Letting Y_0 and Y_1 become the income levels in the two years, and $\overline{Y}_0$, $\overline{Y}_1$ the corresponding per capita levels, and assuming income growth, so $\overline{Y}_1 > \overline{Y}_0$, we can define an income elasticity by

$$e_Y = \frac{\log (\overline{X}_{1Y}/\overline{X}_1)}{\log (\overline{Y}_1/\overline{Y}_0)} = \frac{\log (X_{1Y}/X_1)}{\log (\overline{Y}_1/\overline{Y}_0)}$$

as an average over the period. A demographic elasticity is similarly defined as

$$e_N = \frac{\log (X_1/X_0)}{\log (N_1/N_0)} = 1 + \frac{\log (\overline{X}_1/\overline{X}_0)}{\log (N_1/N_0)}$$

These elasticities then allow total _proportional_ growth in expenditure to be decomposed as

$$\log (X_{1Y}/X_0) = e_y \log (\overline{Y}_1/\overline{Y}_0) + e_N \log (N_1/N_0)$$

It is evident that in order to distinguish the income and composition effects, we have to make projections in which productivity is assumed not to grow. These no-growth projections are of no interest in themselves, since income is virtually certain to continue increasing, but they permit that effect to be separated from the effect of changes in population composition. This approach relates the income effect to the increase in aggregate income, just as the scale effect is related to the increase in aggregate population.

The "composition effect" includes three separate factors characterizing households--size, age of head, and adult/child ratio--but we do not partition the difference between two projections into the parts due to each of these causes. That is because the "causes" are not really independent, all being different aspects of one phenomenon, namely differences in birth rates cumulated over time. Given assumptions about age-specific mortality, the birth rate projection uniquely determines the projected age composition of the population. The only compositional effects that could vary, given a particular population age distribution, are those due to the way people are grouped into households. Here we might have experimented with different assumptions about such groupings so as to isolate a behavioral effect, but there seemed to be little basis for doing so; we chose instead to apply the same behavioral assumptions to all the different structural scenarios. This procedure in effect makes the household size and age distributions also functions of the initial population and the subsequent birth and death rates. Even when we limit the projections in this way, however, the three population scenarios differ enough to generate quite different household structures by the year 2025. In the same way, we did not experiment with different income distributions but let the distributional differences that arise depend only on total income and the demographic differences among scenarios, according to a simple model of how relative incomes are determined. This approach leaves a total compositional effect to study, including the effect of composition on the income distribution, while the income effect refers only to the level of per capita (mean) income.

When two scenarios are compared in a **given** year, making the same
assumption about productivity but different assumptions about population
growth, there is no income effect in the sense defined here: while per
capita income may differ slightly between the two scenarios, that will
only reflect their differences in population composition which affect
labor force participation. In per capita terms, therefore, the entire
difference between expenditure in two scenarios, $\overline{X}_1 - \overline{X}_2$, is attributed
to compositional differences, while the total difference in spending,
$X_1 - X_2$, is decomposed into the terms:

$$X_E \left(\frac{N_1 - N_2}{N_E}\right) \qquad \underline{\text{scale}}: \quad \begin{array}{l}\text{population and spending in scenario E} \\ \text{are used as the basis for comparison}\end{array}$$

$$X_1 - X_2 - X_E \left(\frac{N_1 - N_2}{N_E}\right) \quad \underline{\text{composition}}: \quad \begin{array}{l}\text{the remainder of the} \\ \text{difference}\end{array}$$

A demographic elasticity can still be defined as

$$e_N = \frac{\log (X_1/X_2)}{\log (N_1/N_2)}$$

equal to 1.0 if there is no compositional effect, but no income elasticity
is calculated. This demographic elasticity can be calculated assuming
income growth, or assuming constant productivity.

Summary of Results: Total and Per Capita Spending

Table 1-2 summarizes the projections of total expenditure, by scenario and productivity growth assumptions, in millions of 1971 dollars.
The expenditure sectors are grouped into twelve major categories of spending (expenditures by individual sector are presented in chapter 6). Housing, school and hospital construction are shown in total, and the remaining eleven categories are components of consumption: health and education
include public expenditures, which in the case of education lie outside
PCE. Total spending in the twelve categories studied rises from about
830 billion dollars in 1975 to between 3,106 billion and 3,826 billion by

Table 1-2. Summary of Projections by Expenditure Category, by Scenario
and Year, 1975-2025

(millions of 1971 dollars)

Scenario & category	1975 (base)	1985 static	1985 growth	2000 static	2000 growth	2025 static	2025 growth
Scenario D							
Food (14 sectors)	95326	110695	116845	130385	151422	155403	219774
Clothing (3)	22152	28775	31463	37349	54724	46868	106733
Furnishings (15)	27950	36339	41005	46840	69536	59079	133416
Energy (5)	34588	37766	43700	44275	55021	56897	77716
Vehicles (3)	31059	40966	43997	57058	66335	81945	126311
Public transport (6)	7106	10006	10979	13556	21310	18545	47259
Health care (4)	77300	88700	129200	102900	249200	137100	596700
Education (2)	63000	66621	93119	78771	172131	102431	394425
Trade (2)	156491	200529	219160	259486	363675	354925	805904
Other services (13)	196330	244409	286332	304198	489162	385545	970026
Other goods (66)	77771	45000	48240	54931	82012	69181	194580
Construction (3)	42500	51300	63800	46000	85300	58600	153200
Total consumption plus construction	831573	961106	1127840	1175749	1859828	1526519	3826084
Total PCE	672507	834279	949621	1036287	1557397	1346383	3169059
Scenario E							
Food	95002	100921	113890	127261	145012	147346	187747
Clothing	21982	29059	30212	33262	53958	36413	89726
Furnishings	27803	36623	39791	41676	68604	45420	111897
Energy	34287	36494	42577	40942	51244	47037	63830
Vehicles	30600	41675	42041	46928	66681	53749	107715
Public transport	7051	10140	10614	11562	21858	13341	40946
Health care	76700	86600	139900	97500	253300	119800	558400
Education	64701	62665	107308	69430	183232	71519	335534
Trade	155674	199972	216053	236270	359803	286832	698009
Other services	195682	242373	280676	277545	479477	312797	833520
Other goods	79315	53205	42258	48562	79831	55011	163340
Construction	43350	49300	64100	39100	77100	39400	112800
Total consumption plus construction	832147	949027	1129420	1070038	1840100	1228665	3303464
Total PCE	669388	824538	935412	946598	1532868	1099210	2749630
Scenario F							
Food	95321	111050	113193	126753	143175	143740	171699
Clothing	21936	29228	30535	32619	53357	32140	82568
Furnishings	27582	36789	40035	40784	68033	39837	104133
Energy	34464	35753	41930	38815	48884	41015	56484
Vehicles	30402	42091	44154	45337	66488	41801	99167
Public transport	6993	10220	10744	11256	21685	11375	38543
Health care	76800	87900	142300	96700	266000	112100	542300
Education	64775	60496	105995	61381	167413	56283	269545
Trade	155385	200272	219442	230857	360871	255982	649857
Other services	195355	241201	283310	290756	471921	301073	759017
Other goods	79649	43185	58317	25911	80051	27093	242475
Construction	42900	48600	66600	35800	73300	28800	89900
Total consumption plus construction	831562	946785	1156555	1036969	1821178	1091239	3105688
Total PCE	668941	824874	949660	924418	1529765	987552	2551843

Table 1-2 (continued)

Notes:

Definition of categories (sectors included in each):

Food, drink and tobacco (including food component of meals out of home):
2, 7, 23-34.

Clothing and (leather) footwear: 38, 39, 76

Furnishing and durable goods (excluding machinery and materials for maintenance or construction): 35-37, 40, 45, 123-126, 128, 142, 145-148.

Energy (gasoline, heating oil, natural gas and electricity): 69, 70,
160, 161, 180.

Vehicles (including cycles but not boats, aircraft or mobile homes) and
tires: 72, 133, 139.

Public transport (and travel agents): 151-155, 157.

Health care: 66, 143, 144, 175 (and unallocated public expenditure).

Education: 176, 189.

Trade (wholesale and retail): 163, 164

Other services (including services of owner-occupied dwellings): 158,
162, 166-168, 174, 177.

Other goods (most of which are seldom bought by households): 1, 3, 8,
10, 14, 16, 21, 22, 41, 43, 46, 48, 49, 51-62, 67, 68, 73, 74, 77, 78, 80-83,
87, 90, 93-99, 101-103, 106, 108, 109, 115-117, 119, 120, 122, 127, 129-131,
134, 137, 149, 150, 178, 181.

the end of the projection period: in the absence of productivity growth, the total would reach some 1,091 billion to 1,527 billion dollars.

The results in Table 1-2 are more readily analyzed if they are expressed either in per capita terms or as shares of the total; these are alternative ways of removing the scale effect. Table 1-3 shows per capita spending, in 1971 dollars, in each category for each scenario. Table 1-4 shows expenditure in each category as a percentage of the total, also for each scenario in each year. (This differs slightly from the treatment in chapter 6, where only those sectors are shown which comprise PCE, excluding construction and public consumption.) The shares for almost all categories rise or fall monotonically; the chief exceptions are clothing and household furnishings, where the expenditure share rises until the year 2000 and then declines very slightly; and vehicles, where in the income-growth scenarios the peak share occurs in 1985.

It is clear from table 1-3 that there is a substantial income effect for most categories: expenditure usually rises substantially when income grows, but may grow very little or even decline in the absence of productivity growth. The categories most affected--where per capita spending differs most as income does or does not grow--appear to be services, particularly health care and education but also public transport and the large group of other services. Goods, particularly food, are generally less income-sensitive. Overall, with income growth, per capita spending is projected to increase by a factor of about 1.7 by the year 2000, and about 2.7 by the year 2025; the factor ranges from only about 1.4 for food to 4.8 for health care, over the entire 50-year period.

The most striking result of the comparison of expenditure shares in table 1-4 is the uniformity of the expenditure structure across scenarios. When income grows, the greatest differences among scenarios in the year 2025 are less than 0.5 percent for six of the twelve categories, and only slightly larger for the category of other services. The only substantial differences arise for health care (1.9 percent), education (1.6 percent), housing, school and hospital construction (1.1 percent), and the heterogeneous category of other goods (2.9 percent). The last of these categories, however, consists mostly of sectors for which demand was estimated without allowing for composition effects; the category behaves as a resi-

Table 1-3. Projected Per Capita Expenditures by Category, by Scenario
and Year, 1975-2025

(1971 dollars)

Scenario & category	1975 (base)	1985 static	1985 growth	2000 static	2000 growth	2025 static	2025 growth
Scenario D							
Food	443	454	479	456	530	423	598
Clothing	103	118	129	131	191	128	290
Furnishings	130	149	168	164	243	161	363
Energy	161	155	179	155	192	155	211
Vehicles	144	168	180	200	232	223	344
Public transport	33	41	45	47	75	50	129
Health care	359	364	530	360	871	373	1624
Education	293	273	382	276	601	279	1073
Trade	727	822	898	907	1272	966	2193
Other services	912	1002	1174	1064	1711	1049	2640
Other goods	360	184	198	191	288	188	530
Construction	197	210	262	161	298	159	416
Total spending	3862	3940	4624	4112	6503	4154	10410
Total PCE	3123	3420	3893	3624	5446	3664	8623
Scenario E							
Food	444	471	483	481	548	485	618
Clothing	103	123	128	126	204	120	295
Furnishings	130	155	169	158	259	150	368
Energy	160	155	181	155	194	155	210
Vehicles	143	177	178	177	252	177	355
Public transport	33	43	45	44	83	44	135
Health care	359	367	572	369	958	394	1838
Education	302	266	455	262	694	235	1104
Trade	728	848	917	894	1361	944	2298
Other services	915	1028	1191	1050	1813	1030	2744
Other goods	372	184	202	183	301	180	539
Construction	202	209	272	148	292	130	370
Total spending	3890	4027	4792	4046	6959	4044	10873
Total PCE	3129	3498	3969	3580	5797	3618	9051
Scenario F							
Food	447	481	490	506	571	543	648
Clothing	103	127	132	130	213	121	312
Furnishings	129	159	173	163	271	150	393
Energy	162	155	182	155	195	155	213
Vehicles	142	182	191	181	265	158	374
Public transport	33	44	47	45	87	43	146
Health care	360	381	616	386	1061	423	2047
Education	304	262	459	245	667	213	1018
Trade	728	867	950	921	1440	966	2453
Other services	916	1045	1227	1160	1883	1137	2865
Other goods	372	187	211	102	320	102	915
Construction	201	210	288	143	292	109	339
Total spending	3897	4100	4966	4137	7264	4120	11724
Total PCE	3135	3572	4113	3688	6102	3728	9633

Table 1-4. Percentage Shares of Total Expenditures by Category by
Scenario and Year, 1975-2025

Scenario & category	1975 (base)	Year and Income Assumption					
		1985		2000		2025	
		static	growth	static	growth	static	growth
Scenario D							
Food, drink & tobacco	11.46	11.52	10.36	11.09	8.14	10.18	5.74
Clothing & footwear	2.66	2.99	2.79	3.18	2.94	3.07	2.79
Furnishings & dur- ables	3.36	3.78	3.64	3.98	3.74	3.87	3.49
Energy	4.16	3.93	3.87	3.77	2.96	3.72	2.03
Vehicles & tires	3.73	4.26	3.90	4.85	3.57	5.37	3.30
Public transport	0.85	1.04	0.97	1.15	1.15	1.21	1.24
Health care	9.30	9.23	11.46	8.75	13.40	8.99	15.60
Education	7.58	6.93	8.26	6.70	9.26	6.71	10.31
Trade	18.82	20.86	19.43	22.07	19.55	23.21	21.06
Other services	23.61	25.43	25.39	25.87	26.30	25.21	25.35
Other goods	9.35	4.68	4.28	4.67	4.41	4.52	5.09
Construction	5.11	5.34	5.66	3.91	4.59	3.84	4.00
Total PCE	80.87	86.80	84.20	88.14	83.74	88.20	82.83
Scenario E							
Food, drink & tobacco	11.42	11.69	10.08	11.89	7.88	11.99	5.68
Clothing & footwear	2.64	3.06	2.68	3.11	2.93	2.96	2.72
Furnishings & dur- ables	3.34	3.86	3.52	3.89	3.73	3.70	3.39
Energy	4.12	3.85	3.77	3.83	2.78	3.83	1.93
Vehicles & tires	3.68	4.39	3.72	4.39	3.62	4.37	3.26
Public transport	0.85	1.07	0.94	1.06	1.19	1.09	1.24
Health care	9.22	9.12	12.39	9.11	13.77	9.75	16.90
Education	7.78	6.61	9.50	6.49	9.96	5.82	10.16
Trade	18.71	21.07	19.13	22.08	19.55	23.35	21.13
Other services	23.52	25.54	24.85	25.94	26.06	25.46	25.23
Other goods	9.53	5.61	3.74	4.54	4.34	4.48	4.94
Construction	5.20	5.19	5.68	3.65	4.19	3.21	3.41
Total PCE	80.44	86.88	82.82	88.46	83.30	89.46	83.23
Scenario F							
Food, drink & tobacco	11.46	11.73	9.79	12.22	7.86	13.17	5.53
Clothing & footwear	2.64	3.09	2.64	3.15	2.93	2.95	2.66
Furnishings & dur- ables	3.32	3.89	3.46	3.93	3.74	3.65	3.35
Energy	4.14	3.78	3.63	3.74	2.68	3.76	1.82
Vehicles & tires	3.66	4.45	3.82	4.37	3.65	3.83	3.19
Public transport	0.84	1.08	0.93	1.09	1.19	1.04	1.24
Health care	9.24	9.28	12.30	9.33	14.61	10.27	17.46
Education	7.79	6.39	9.16	5.92	9.19	5.16	8.68
Trade	18.69	21.15	18.97	22.26	19.82	23.46	20.92
Other services	23.49	25.48	24.50	28.04	25.91	27.59	24.44
Other goods	9.58	4.56	5.04	2.50	4.40	2.48	7.82
Construction	5.16	5.13	5.75	3.45	4.02	2.64	2.89
Total PCE	80.44	87.12	82.11	89.15	84.00	90.50	82.17

dual in total spending. Only for health, education and construction does
the composition of the population greatly affect the structure of spending, and the sum of the three categories is nearly constant. To a first
approximation, then, slower population growth out to 2025 simply means
greater spending on health and compensating reductions for education and
construction.

There would be less uniformity if population were assumed to grow
while productivity and income did not, and the largest differences would
then appear in other categories. Scenarios D and F would differ in 2025
by 3.0 percent for food, 2.4 percent for other services, 2.0 percent for
other goods, and 1.0 percent or more for vehicles, health care, education
and construction. Evidently, the demographic differences visible in this
comparison are reduced rather than accentuated by income growth. Thus,
scale and income effects appear to account for most of the difference
among scenarios--at least at this level of aggregation--with compositional
effects being important in only a few sectors.

When 1975 is compared to 2025, assuming income growth, the largest
absolute shifts are in food (-5.7 to 5.9 percent, depending on the scenario), health care (+7.7 percent, using scenario E for comparison), and
education (2.7 percent in scenario D, 2.4 in scenario E). There are also
large relative changes for household energy use (-2.2 percent), public
transport (+0.45 percent) and construction (as much as -2.3 percent). In
any scenario expenditure shifts away from goods and toward services, but
it should be noted that health care and education grow much faster than
other services. When intermediate years are compared, it is evident that,
with the exceptions noted, the changes in expenditure shares are roughly
constant through time: thus the largest changes usually occur in the
final 25-year period rather than in the previous shorter intervals.

Scale, Composition and Income Effects

Growth of expenditure can be decomposed into these three effects
over any interval of time, but we consider here only the three periods
beginning in 1975 and extending to each of the other benchmark years.
Table 1-5 presents the decomposition, for eleven of the twelve categories

Table 1-5. Percentage Decomposition of Total Expenditure Growth, by
Expenditure Category, by Scenario and Period

Category & Scenario		1985			2000			2025		
		SCA	COM	INC	SCA	COM	INC	SCA	COM	INC.
Total Spending	D	36.8	8.7	54.4	27.7	7.2	65.1	20.7	3.7	75.6
(12 categories)	E	31.4	11.9	56.7	20.8	4.4	74.8	15.1	2.0	82.9
	F	23.7	16.2	60.1	15.6	6.4	78.0	9.7	2.9	87.4
Food, drink &	D	58.9	12.5	28.6	55.8	6.7	37.5	54.1	-5.8	51.7
tobacco	E	51.2	33.1	15.7	44.8	19.7	35.5	43.0	13.4	43.6
	F	43.8	44.2	12.0	34.8	30.9	34.3	30.1	33.3	36.6
Clothing &	D	31.6	39.5	28.9	22.3	24.4	53.3	18.5	10.7	70.8
footwear	E	27.2	58.8	14.0	16.2	19.1	64.7	13.6	7.7	78.7
	F	21.0	63.8	15.2	12.2	21.8	66.0	8.7	8.1	83.2
Furnishings &	D	28.4	35.9	35.9	22.1	23.3	54.6	18.7	10.8	70.5
durables	E	23.6	50.0	26.4	16.1	17.9	66.0	13.9	7.0	79.1
	F	18.2	55.7	26.1	11.9	20.7	67.4	8.7	7.3	84.0
Energy	D	50.4	-15.6	65.1	55.5	-8.1	52.6	56.7	-4.9	48.3
	E	42.1	-15.5	73.4	47.7	-8.5	60.8	48.8	-5.6	56.8
	F	37.9	-20.7	82.7	41.8	-11.6	69.8	37.8	-8.0	70.2
Vehicles &	D	31.9	44.7	23.4	28.9	44.8	26.3	23.0	30.4	46.6
tires	E	27.2	69.6	3.2	20.0	25.3	54.7	16.7	13.3	70.0
	F	18.2	66.8	15.0	14.7	26.7	58.6	10.7	5.9	83.4
Public transport	D	24.4	50.5	25.1	16.4	27.0	54.6	12.5	16.0	71.5
	E	20.1	66.6	13.3	11.2	19.3	69.5	8.7	9.9	81.4
	F	15.3	70.7	14.0	8.3	20.7	71.0	5.4	8.5	86.1
Health care	D	19.8	2.2	78.0	14.8	0.1	85.1	10.5	1.0	88.5
	E	13.4	3.6	83.0	10.3	1.5	88.2	6.7	2.2	91.5
	F	9.6	7.3	83.1	7.1	3.4	89.5	4.0	3.6	92.4
Education	D	27.8	-15.8	88.0	18.9	-4.5	85.5	13.4	-1.5	88.1
	E	15.5	-20.2	104.8	12.9	-8.9	96.0	10.0	-7.5	97.5
	F	12.9	-23.3	110.4	11.0	-14.3	103.3	7.6	-11.8	104.1
Trade	D	33.2	37.1	29.7	24.8	24.9	50.3	17.0	13.5	69.4
	E	26.2	47.1	26.6	18.0	21.5	60.5	12.1	12.1	75.8
	F	19.9	50.1	29.9	13.2	23.5	63.3	7.6	12.8	79.7
Other services	D	29.0	24.4	46.6	22.0	14.8	63.2	17.9	6.5	75.5
	E	23.4	31.5	45.1	16.3	12.6	71.2	12.9	5.5	81.6
	F	18.3	33.9	47.9	12.3	22.1	65.5	8.4	10.4	81.4
Construction	D	26.5	14.8	58.7	32.6	-24.4	91.8	27.1	-12.6	85.5
	E	21.2	7.6	71.2	30.2	-42.6	112.4	26.2	-31.8	105.6
	F	14.9	9.2	75.9	24.7	-48.1	123.4	22.0	-52.0	130.0
Total PCE	D	32.2	17.1	50.7	24.9	13.2	61.9	19.0	6.7	74.3
	E	25.6	24.4	50.0	18.3	10.1	71.6	13.5	5.4	81.1
	F	19.6	27.4	53.0	13.6	12.2	74.2	8.6	6.3	85.1

Table 1-5 (continued)

Alternative classification by type of estimating equation

Category & scenario		1985			Period: 1975 to 2000			2025		
		SCA	COM	INC	SCA	COM	INC	SCA	COM	INC
Demographically sensitive sectors*	D	42.3	11.1	46.6	28.1	12.5	59.4	20.5	6.2	73.3
	E	34.3	22.1	43.6	20.4	9.5	70.1	14.6	5.0	80.4
	F	25.7	27.9	46.4	15.1	12.6	72.3	9.2	6.6	84.2
Time-series sectors**	D	34.4	0.0	65.6	27.2	0.0	72.8	21.2	0.0	78.8
	E	27.2	0.0	72.8	21.2	0.0	78.8	15.5	0.0	84.5
	F	21.7	0.0	72.3	16.4	0.0	83.6	10.3	0.0	89.7

SCA Scale Effect: percentage of expenditure growth attributed to increase in population size.

COM Composition Effect: percentage of expenditure growth attributed to change in population composition (including income redistribution).

INC Income Effect: percentage of expenditure growth attributed to increase in labor productivity.

* 82 PCE sectors for which demand equations were estimated from the consumer expenditure survey, or from other demographic data.

** 51 PCE sectors for which Ridker and Watson's aggregate time-series demand equations were used. "Composition" effects for these sectors reflect time trends in the demand equations.

Because the category "other goods" show negative growth for some periods and scenarios, no decomposition analysis is shown.

and their sum, and also for two subtotals--the 82 sectors for which we estimated new, demographically-sensitive demand functions, and the remaining 51 sectors for which we used Ridker and Watson's aggregate time-series equations. For this last group, there is no true composition effect but only scale and income factors. What appear as composition effects in the sectors estimated from time-series data are actually due to time trends or to a residual not accounted for by the scale and income effects. In every scenario and period, the total demographic effect (scale plus composition) is larger for the former than for the latter group of sectors, and the income effect is smaller. This is partly because introducing demographic detail makes it easier to discern demographic effects, in any one sector, but mostly it is due to the fact that demographic effects really are more important in the first group; equations estimated for sectors in the second group often showed no significant effects of population composition.

Through time, demographic effects usually decline in relative importance, because population growth slows down while productivity grows at essentially constant rates. Over the entire 50-year projection period, population change (both in size and composition) accounts for one-fourth or less of total expenditure growth, although through the first decade it is roughly equal to income in importance. Out to 2025, the growth of income provides less than half the growth of expenditure only for food, drink and tobacco (all scenarios), and energy and vehicles (scenario D only). Where both the scale and composition components are positive, they usually decline together; the scale effect is more important, except for the categories of clothing, furnishings, vehicles, and other services (in scenario E). The first three of these are categories in which spending is quite sensitive to age and household composition; furniture and vehicles were also adjusted in a way that may have exaggerated the compositional term.

Across scenarios, demographic effects are usually greatest when population grows most rapidly, being largest in scenario D and smallest in F. (The only exception to this pattern is food spending; incomes are higher in scenario F, but food expenditure tends to saturation as income rises.) It is mostly the scale effect which produces this association; the composition effect is often largest when population grows most slowly. This reflects the fact that the population changes most from its <u>present</u> composition in scenario F, when fertility declines most. Scenario D is closest to a continuation of the present composition, so it involves the least compositional change.

For many categories, composition is never very important as a determinant of spending; scale effects are significant in the short run, while beyond the year 2000 income growth accounts for nearly all of expenditure change. The exceptions to this pattern are food, for which income makes less difference; energy, for which price changes and projected conservation reduce the importance of income; and two categories for which changes in population composition act to <u>reduce</u> expenditure (that is, the composition effect is negative). These are education (for all scenarios and periods), and construction (for all scenarios after 1985). These are pure compositional effects, not caused by changes of prices or tastes but due to changes in the numbers of children and in the size distribution of households. They lead to a net negative demographic effect in scenario F, and for construction, in E as well. Health care might be expected to show important compositional effects, since such expenditures are concentrated among the aged, but the effect is small compared to that of income. Also, the older the population the lower is the birth rate, and birth-related expenses are a large share of medical care.

Finally, income and demographic effects can also be analyzed through time by means of elasticities. These are presented in table 1-6. For this purpose, "income" is defined as per capita disposable income, the instrumental variable in the projections. Note that the income elasticity e_Y is defined by the change which income growth would make in a future year, in per capita terms, so that it is independent of population growth, but not of population composition. The demographic elasticity e_N is defined by the growth in expenditure that would occur because of demographic change, in the absence of income growth. Either type of elasticity can be affected by changes in the composition of the population; in the case of e_N, this is the only reason for the elasticity to differ from 1.0.

For total spending and for several of its components, both elasticities tend toward 1.0 through time, in all three scenarios. However, even

Table 1-6. Income and Demographic Elasticities by Expenditure Catetory, by Scenario and Period

Category and Scenario		Period: 1975 to:					
		1985		2000		2025	
		e_Y	e_N	e_Y	e_N	e_Y	e_N
Food, drink and tobacco	D	0.227	1.197	0.247	1.102	0.361	0.914
	E	0.094	1.609	0.194	1.378	0.235	1.252
	F	0.056	1.928	0.166	1.766	0.160	1.900
Clothing & footwear	D	0.377	2.090	0.619	1.847	0.853	1.406
	E	0.149	2.831	0.718	1.951	0.873	1.436
	F	0.117	3.652	0.677	2.438	0.858	1.745
Furnishings & durables	D	0.508	2.093	0.646	1.819	0.848	1.400
	E	0.324	2.814	0.737	1.920	0.871	1.408
	F	0.257	3.648	0.697	2.445	0.873	1.697
Energy	D	0.579	0.704	0.339	0.870	0.311	0.930
	E	0.497	0.643	0.313	0.837	0.282	0.901
	F	0.408	0.465	0.298	0.738	0.271	0.806
Vehicles & tires	D	0.292	2.236	0.244	2.158	0.452	1.818
	E	0.021	3.200	0.526	2.006	0.675	1.608
	F	0.147	4.142	0.523	2.499	0.781	1.494
Public transport	D	0.394	2.740	0.768	2.246	0.989	1.777
	E	0.171	3.731	0.946	2.357	1.088	1.820
	F	0.201	4.643	0.904	2.916	1.108	2.224
Health care	D	1.591	1.111	1.452	1.010	1.535	1.072
	E	1.664	1.227	1.422	1.130	1.495	1.265
	F	1.461	1.718	1.386	1.431	1.429	1.746
Education	D	1.418	0.433	1.284	0.789	1.406	0.908
	E	2.017	-0.309	1.446	0.330	1.500	0.285
	F	1.706	-0.883	1.375	-0.333	1.420	-0.645
Trade	D	0.353	1.984	0.527	1.781	0.818	1.531
	E	0.250	1.988	0.586	1.959	0.822	1.736
	F	0.234	3.212	0.577	2.459	0.788	2.311
Other services	D	0.628	1.752	0.741	1.542	0.921	1.261
	E	0.473	2.206	0.761	1.649	0.906	1.336
	F	0.412	2.669	0.626	2.470	0.782	2.002
Construction	D	0.937	1.512	1.012	0.289	1.006	0.599
	E	0.988	1.351	1.013	-0.468	1.018	-0.257
	F	0.961	1.555	0.979	-1.103	1.028	-1.829

Table 1-6 (continued)

Category and Scenario		1985 e_Y	1985 e_N	2000 e_Y	2000 e_N	2025 e_Y	2025 e_N
Total consumption	D	0.635	1.160	0.715	1.221	0.917	1.136
& construction	E	0.561	1.357	0.755	1.186	0.914	1.111
	F	0.512	1.643	0.728	1.369	0.884	1.257
Total PCE	D	0.514	1.728	0.635	1.524	0.854	1.299
	E	0.407	2.150	0.671	1.635	0.847	1.414
	F	0.360	2.653	0.651	2.007	0.802	1.801

Period: 1975 to:

The category "Other Goods" is not analyzed because of negative growth in some periods and scenarios.

over the entire half-century projection period, several of the income elasticities remain substantially different from one: those for food are quite low, those for energy actually <u>decline</u> through time, and those for health care and education remain at 1.4 or higher. The demographic elasticities show still more variation from 1.0; although they move toward this value for nearly all categories, there are still some significant compositional effects, particularly in scenario F. For education and for construction, the elasticities with respect to population growth become negative, reflecting the negative components of growth calculated in table 1-5.

An elasticity can also be used to compare two scenarios in the same year: a value of 1.0 means there is no composition effect, a higher value means that the composition effect raises spending faster than population increases between the two scenarios, and a lower value indicates that the composition and scale effects are opposed. If the elasticity is negative, differences in population structure outweigh scale differences, spending being higher when population is lower.

Table 1-7 compares scenarios D and F to E, assuming productivity growth, in 1985, 2000 and 2025. Beyond 1985 nearly all elasticities are positive, and they show some tendency to converge to 1.0, indicating again that the composition effect is reduced over time. However, even in 2025 the elasticities for education (in scenario F) and construction (in both

Table 1-7. Demographic Elasticities, Scenarios D and F Compared to E,
            by Expenditure Category and Year)
            (Assuming Income Growth)

Category	Year Scenario:	1985 D	1985 F	2000 D	2000 F	2025 D	2025 F
Food, drink & tobacco		0.746	0.259	0.552	0.239	0.827	0.652
Clothing & footwear		1.182	-0.518	0.180	0.210	0.912	0.607
Furnishings & durables		0.875	-0.298	0.172	0.157	0.924	0.525
Energy		0.758	0.746	0.908	0.883	1.034	0.892
Vehicles & tires		1.324	-2.389	-0.066	0.054	0.837	0.603
Public transport		0.965	-0.553	-0.324	0.149	0.753	0.441
Health care		-1.257	-2.602	-0.208	-0.917	0.349	0.214
Education		-4.130	0.600	-0.798	1.692	0.850	1.598
Trade		0.416	-0.758	0.137	-0.056	0.755	0.522
Other services		0.581	-0.455	0.255	0.298	0.797	0.683
Other goods		3.855	-15.697	0.344	-0.052	0.919	-2.883
Construction		-0.137	-1.864	1.291	0.947	1.608	1.656
Total consumption plus construction		-0.041	-1.157	0.136	0.194	0.772	0.451
Total PCE		0.439	-0.737	0.203	0.038	0.746	0.545

scenarios) are far above one, and those for health care remain far below it.
In the remaining categories, composition effects nearly always offset
slightly the differences due to population size alone: the difference in
demand between two possible population growth paths is usually less than
we would expect from scale differences.

The shift away from goods and toward services follows, in our projections, directly from the income-responsiveness of the different consumption
functions. We make no assumptions about differential rates of productivity
growth in the production of goods and services, and do not allow for changes
in relative prices. One might suspect, however, that if services take a
larger share of the consumer budget in the future, that will be at least
partly because productivity will grow more slowly in services and their
unit costs will in fact rise, while demand for services will be relatively

price-inelastic. Since for most sectors we did not project unit labor
requirements and productivity, we could not base projections of expenditure on that assumption; moreover, the CES data provide no basis for estimating responses to price changes. In the case of education, we did project personnel requirements directly, and then determined costs by assuming a rate of increase of educational salaries. If such salaries were
held constant as a way of holding down unit costs--or if, alternatively,
faculty-student ratios could fall so as to offset rising wages, reflecting
more productivity per teacher--expenditures would be lower. A comparison
of projections based on this assumption with our standard projections assuming no productivity growth but rising wages, shows the effect, for this
one category of spending, of differential productivity growth.

The two projections are compared extensively in chapter 3. Here we
note only that the constant-cost-per student projections would lead to per
capita spending, in the year 2025, of only $560-$660, instead of the
$1,000-$1,100 shown in table 1-3. Education would absorb between five and
seven percent of total (consumption plus construction) spending, rather
than the nine or ten percent shown in table 1-4. The lower expenditures
would appear to depend less on income growth and more on demographic factors, both scale and composition effects being increased in magnitude for
all three scenarios. (Whenever the scale effect is more than offset by
the negative composition effect, the income effect would, in consequence,
increase in magnitude also.) Finally, income elasticities would be about
0.8 to 0.9 rather than, as in table 1-6, between 1.4 and 2.0. The difference of 0.6 to 1.1 can be attributed to unit cost increases almost entirely, since enrollments at most school levels are assumed not to vary with
costs or prices. This exercise suggests that in fact a large part of the
growth projected for service expenditures will originate in increases in
unit cost (or in declining relative productivity) rather than in increases
in quantity. The generalization is complicated, however, by the difficulty of defining quantity and quality for such services as education and
health: a constant teacher-student or doctor-patient ratio may be associated with rising productivity in the quality of instruction or medical
care.

Summary: How Much Does Demographic Growth Matter?

The importance of demographic change for the future level and structure of final demand can be assessed in two different ways. One is to
analyze the demographic sensitivity of the expenditure functions for individual sectors, both directly and by comparison to aggregate equations
which take no account of compositional effects. The detailed discussion
of the estimating equations appears in chapters 2 through 5; here we present a brief summary. The other approach is to decompose the projection
results, as in the analysis presented in tables 1-5 through 1-7. So far,
we have conducted this sort of analysis only at the level of eleven or
twelve major categories of spending, leaving the more detailed results by
sector for chapter 6.

Several generalizations can be drawn from the individual demand equations estimated from the consumer expenditure survey data (chapter 2).
First, a large number of sectors show no significant demographic effects,
so they can be adequately represented as functions of income and total
population alone, justifying the aggregate approach taken by previous
investigators. Most of these sectors are, however, quite small; 51 of
them together sum to only 18.5 percent of personal consumption spending
in the base year, whereas the same number of sectors estimated with demographic effects sum to 25.2 percent of PCE. Another 25 sectors, accounting for 22.3 percent of PCE, also show demographic effects but require
some form of adjustment to override effects attributed to income or to
introduce reasonable time trends. As table 1-1 indicates, demographic
effects were taken into account using different data and models for another six sectors including 11.2 percent of PCE. Wholesale and retail
trade, which provide the remaining 22.8 percent, were not separately
analyzed.

Second, most consumption sectors approach relative saturation, tending to a share of the budget of one percent or less. Higher shares characterize only housing services, durables, vehicles, private education and
a number of luxury categories associated either with travel or with services such as credit, insurance and real estate brokers. The income coefficients are sometimes demographically sensitive, differing across house-

hold size classes. This means that changes in the household size distribution will affect the relation of expenditure to income. The chief size distinction is between one- and many-person households; single individuals tend to have systematically different spending habits, especially for food. This effect is important because with slower population growth in the future, such households are expected to increase more rapidly than any other kind.

Third, age differences are not generally very significant once household size is taken into account; there are, however, fairly large effects for many food sectors and for some services. Because the effects of age on spending are not always monotonic, it is not possible to describe an overall shift in consumption as the population gets older. Most of the change associated with "aging" seems actually to be due to changes in the household size distribution. The exceptions are sectors in which demand does not appear to depend on household structure. For education (chapter 3), slower population growth means a declining demand, apart from income effects. For health (chapter 4), slower growth does not have a large net effect, because while demand rises sharply for the elderly, expenditure is also high at birth and for the very young. A shift from one end to the other of the age distribution therefore changes the composition of medical spending more than the level. Age also appears to affect housing demand, since the preference for houses versus apartments is highest in the middle years (chapter 5). Much of this effect, however, is due to the presence of children in those years, and the housing model does not take household composition explicitly into account, but relies only on family size and age of the head as proxy indicators of composition.

Fourth, significant effects of household composition tend to appear for the same sectors that show substantial age effects. The balance of adults and children in a household is particularly important for most food sectors, alcohol and tobacco, local travel, lodging places, and a few other sectors, many of which are associated with home-owning. Household compositional effects are not considered for health, education and housing, these sectors being based only on the age distribution of all people in the population. Among the sectors showing some compositional effect, there is usually a sharp difference between households consisting only of adults and those with at least one child, whereas the exact number of children does not matter, once there are any at all. This indicates that a reduction in the number of children spread uniformly across households would affect the structure of spending much less than an equivalent change which increased the rate of childlessness.

Taken together, these findings imply that disaggregated demographic effects do matter for the structure of final demand, but such effects are not invariably large and they contain few real surprises. The structure of the population is important for these sectors where one might expect it to be so--health, education, housing, food, alcohol, and tobacco in particular. Most household goods and services are not very demographically sensitive except for life-cycle effects, although there are substantial effects on some sectors associated with homeowning or with travel and recreation. Such demographic effects as are found in our demand equations may, of course, be either reinforced or offset when combined with the projected household structure, and apart from that they may, over the next half-century, be insubstantial relative to the changes caused by income growth.

Since ours are the first projections of the composition of demand to take account of the detailed structure of the population, we can find out how much difference this detail makes by comparing our results to those obtained by previous investigators using aggregate demand functions estimated from time-series data. Our projections share the same definition of scenarios--the same assumptions about population and income growth--as those of Ridker and Watson, so theirs offer the best basis for comparison. (This basis is reinforced by our use of their aggregate projections for a number of sectors.) The comparison is limited to the more probable case, in which labor productivity is assumed to grow.

The comparison is summarized in table 1-8, which shows the frequency distribution and the mean absolute value of the percentage differences between the two types of projection, for the 82 individual sectors for which equations estimated at least in part from the CES were used. (As explained above, such equations showed essentially no demographic effect for the remaining 51 sectors.)

Table 1-8. Summary of Percentage Differences Between Aggregated and
Disaggregated Projections, by Scenario and Year, 1975-2025
(with Productivity Growth)

Year	Scenario	Mean absolute difference	No. of Sectors with Percentage Difference:				
			Under 10	10-30	30-50	50-100	Over 100
1975	D	8.87	45	33	3	1	0
	E	8.63	46	31	4	1	0
	F	8.67	45	32	4	1	0
1985	D	14.26	33	30	14	4	1
	E	14.83	32	33	12	4	1
	F	17.40	31	40	8	2	1
2000	D	20.65	18	37	15	7	5
	E	21.24	21	33	15	8	5
	F	22.34	18	38	13	9	4
2025	D	29.46	12	25	25	9	11
	E	30.07	14	25	21	15	7
	F	30.88	21	18	23	16	4

Mean absolute difference calculated as $\frac{1}{82} \Sigma \mid \% \text{ Diff} \mid$ (% PCE, Aggregate), where the sum is over the 82 sectors which differ between the aggregated and the disaggregated estimates. Absolute values of differences are weighted by shares of PCE. Sectors with less than 0.005% PCE are excluded. Disaggregated projections are from Chapter 6; aggregated projections are those of Ridker and Watson (1980).

Four results are immediately apparent: (1) in any one year, there is not much difference among the three scenarios, except perhaps in 2025, when D shows many more large discrepancies than E or F; (2) as is to be expected, the differences between the two types of projections, for a given scenario, grow wider over time; (3) for only a few sectors are there very large discrepancies; and (4) on average the difference between two sectoral projections rises to about 30 percent over the half-century. It appears that the introduction of more demographically sensitive functions does not radically change the way expenditure is allocated among sectors: nonetheless, by the end of the period about half of the sectors for which new equations were estimated, show differences of 30 percent or more. Thus our findings do not simply replicate those of earlier analysis, and in fact show on average a substantial difference.

Two further observations on this comparison are in order. The first is that large discrepancies would be more frequent, but for the adjustments we made to many sectoral equations. The cross-section survey data alone, unmodified by exogenous estimates of demand level or growth, would show a very different allocation of expenditure in some important sectors. Demographic effects therefore might be considerably larger than we show. The second observation is that much of the remaining difference between the two sets of projections is due to different assumptions about trends, rather than to the introduction of more detailed demographic effects. For example, there are very large discrepancies for sectors 52 (newspapers), 174 (motion pictures and amusements) and 177 (postal services): in Ridker and Watson's projections, these are all assumed to be largely displaced by electronic substitutes, but since there is no evidence for such substitution in the survey data, our equations show continued growth in all three sectors. Similarly, the time-series equations include the assumption that fresh fruits and vegetables (sector 7) will effectively disappear from the market in favor of canned and frozen foods (sector 25), but there is no evidence for this in the cross-section. Drugs (sector 66) show much higher expenditure in our equations because the introduction of public medical programs shifts costs away from consumers. Fabrics for home sewing (sector 35) were assumed to go out of use in Ridker and Watson's projections, but not in ours. Assumptions about future behavior

also affect projections for sectors 29 (confectionary products), 32 (fats
and oils), 80 (pottery), 146 (watches and clocks) and 147 (silver and
jewelry). These are all sectors to which we applied adjustments to offset
the growth that would otherwise occur with increased income.

We conclude this review with a summary of the projection results,
which depend on the assumptions about population and productivity growth,
and the endogenous changes in income distribution and household composition, as well as the characteristics of the individual sector equations.

First, so far as demographic change is concerned, scale effects are
dominant: to a good first approximation, the difference between any two
scenarios in the same year is simply one of size, with the level of income
per person and the structure of spending almost identical. There are no
visible thresholds or discontinuities to produce marked differences in per
capita spending.

Second, such differences as arise through time in per capita terms
usually owe more to income differences than to differences in population
composition. This is particularly true over long intervals, and the relative importance of income rises as the interval is longer or we approach
the end of the projection period. Even for sectors such as foods where
considerable compositional effects might be expected, differences in income are still important. When no income growth is assumed, of course,
demographic effects are the only ones present, and the scenarios differ
much more as to the structure of demand. We can generalize to the conclusion that demographically-related shifts in demand will be important if
income grows slowly, but not if it grows rapidly; in the latter case most
changes in spending can be accounted for by income and population growth
alone.

The distribution of income among households retains the same general
shape through time and across scenarios, as described in chapter 2, but is
subject to stretching and compression as the population structure changes.
A decline in the number of households with young heads, as the population
ages, makes the distribution more equal, but it is made less equal by
increased numbers of the elderly. A reduced size variation pushes toward
greater equality, but none of these effects is very large. Income growth
overshadows any redistribution in our projections.

Third, it happens that some categories which are undoubtedly sensitive to population composition, particularly health and education, are also relative luxuries with pronounced income effects. (These are sometimes manifested through price changes.) As a result, the dominance of income effects is greater than it would be if there were no such association.

Fourth, where compositional effects are pronounced, they owe most to differences in age structure rather than to variations in the distribution of household sizes or in the distribution of children among households. The three scenarios differ more in age structure than in any other respect except scale, and because certain expenditures—durables, health care, education, insurance—are concentrated in particular age groups, these differences can be significant.

Fifth, all the projections show a shift from goods toward services, partly because there is saturation for a number of goods but not for any of the services. However, this shift is not very large even over time, and it is virtually identical across scenarios. It does not depend on a particular population structure, and owes more to income growth than to anything else.

Finally, it should be remembered that so far as any particular sector or category is concerned, the largest determinant of future consumption may not be any of the factors whose effects we have tried to measure, but the combination of tastes and technology which decide how particular wants are to be satisfied. Very large shifts among sectors are conceivable—as between print and electronic media, or among different modes of transport or different fuels—but they are likely to affect rather small shares of total consumption.

It would not be fair to conclude from this summary that the structure of demand is impervious to changes in how the economy grows, or that nothing makes any difference. Size obviously does matter; and even small differences in structure may be of some importance. But what principally changes the composition of spending is the growth of income per person, and that growth is largely independent of the population path the society takes. A difference of 100 million people in the United States in the year 2025 would, of course, lead to large differences in the level of spending on nearly every component of demand, but it does not appear that it would also lead to a major restructuring of demand in the economy.

Chapter 2

PERSONAL CONSUMPTION EXPENDITURE:
ESTIMATION FROM THE CES DATA

Except for three major categories--education, health, and housing,
school and hospital construction--all the components of final demand
analyzed here are entirely included in personal consumption expenditure.
In this chapter, we discuss how projections of demand were made for those
sectors, and provide an initial valuation of the consumption functions
used, based on the 1972-73 Consumer Expenditure Survey.

This chapter is organized as follows. First, we describe the data
used and steps by which they were prepared for analysis. Then we discuss
the form of the consumption functions adopted and the way that income
and demographic effects were taken into account. The two subsequent sec-
tions describe the projections of the number and type of households in
each future year, and the average income of each type of household as a
function of average income in the economy and its demographic composition.

The reader who wants only to know how, in general, the projections
were made need go no further. The rest of the chapter is devoted to an
examination of the coefficients of the expenditure equations, so as to
determine how much, and in what ways, they are demographically sensitive.

Data Preparation

The bulk of personal consumption expenditure is in sectors for which
consumption equations were estimated using the Census Bureau/Bureau of
Labor Statistics 1972-73 consumer expenditure survey (CES). Before equations could be estimated, these data--referring to a subset of 20,000
households of the approximately 68,000 interviewed in the survey--had to
pass through a series of transformations. These transformations take
data in current dollars at retail prices, for categories defined in the

survey and with either a weekly or an annual period of reference as the
data come from the diary survey or from the separate interview survey, and
convert them to producer prices in 1971 dollars at annual rates, for the
sectors defined in the INFORUM model and described in table 1-1. Five
steps are required: deflation to 1971 prices, annualizing (for the diary
data only), removal of sales and excise taxes, removal of wholesale and
retail trade margins, and reclassification into INFORUM sectors.

Deflation

For the diary data, which refer to a one or two-week period between
mid-1972 and mid-1974, monthly indexes were computed for each of the 24
months from July 1972 to June 1974, using the average for 1971 as the base.
For each survey expenditure category, the index was an element (or an
average of elements) of the consumer price index (CPI), and the index used
for a given household (survey observation) was that for the month in which
its diary was completed. An index can therefore be wrong by at most one
half-month. The diary expenditure categories are, for the most part, ones
for which the appropriate elements of the CPI are easily identified.

The interview data refer to total annual expenditure for either 1972
or 1973, so only two indexes, both using the 1971 annual averages as the
base, needed to be defined for each category. For most categories it was
easy to find appropriate elements of the CPI, but there are some expenditures which the CPI does not cover or does not match closely. Where possible, we made indices for these categories, relying on trade journals or
similar sources for a notion of price changes in 1971-73. Where no better
index could be found, the total CPI was used.

Annualizing was required only for the diary data, the interview data
being already on annual basis. There are two reasons why this is not
simply a matter of multiplying weekly figures by 52. The first is that
most households spend some time away from home during the year, usually
on vacation, so that the buying pattern reflected in the diary information is not continued all year. The BLS estimated the average number of
weeks away from home in each survey year, and we used 52 less this number
as our annualizing factor. In fact, time away from home undoubtedly
varies among households—the rich travel more than the poor, for example—

so the adjustment ought to differ by income level and family composition. In the absence of information on this variation, we applied the same adjustment to all households, allowing only for a difference between 1972 and 1973 (the oil price increase and gasoline shortages in the latter year curtailed travel somewhat).

The second reason for a more complicated adjustment is that households occasionally stock up on certain foods and other items of frequent purchase—canned or frozen foods, dry cereals, paper products—so that their annual purchases would be much less than 52 times the expenditure recorded in a week. In principle, such bulk purchases are offset, at least so far as mean or total consumption is concerned, by households which stocked up at another time and recorded no purchases in the survey. This compensation can, however, leave regression coefficients biased if bulk purchases are more common among some types of households than others. We therefore excluded observations of extremely high expenditures, eliminating between one in a thousand and one in ten thousand observations for each of the affected sectors.

Removal of Taxes

The survey data report expenditure at consumer prices, including any federal, state and local sales or excise taxes. As many as three layers of tax have therefore to be stripped away to get to pre-tax prices. 1972 tax rates by category at each level of government were taken from Significant Features of Fiscal Federalism, 1976-77 Edition (ACIR, 1977). Sales taxes vary by state and to some extent locally, but regional patterns are apparent. Households are identified in the CES only by region (North East, North Central, West and South), not by state, so weighted regional averages are also calculated for state and local tax rates (local sales taxes were considered only if they are uniform within a state). Each state's total personal income was used to weight its tax rates in arriving at the average for a region. These regional rates, by major categories, are shown in table 2-1. Federal excise taxes for major items such as alcoholic beverages, cigarettes, gasoline and automobiles were added uniformly to the regional rates; state excise taxes on alcohol and tobacco were also included. Pretax expenditure in a category was then estimated

Table 2-1. Regional Sales Tax Rates (percentages), 1972

Category	North East	North Central	West	South
General Sales	5.81	3.75	4.35	4.17
Food	0.0	1.42	1.04	1.99
Motor Vehicles	5.81	3.72	4.34	2.96
Admissions	0.81	1.12	0.98	1.67
Restaurants	6.40	3.75	4.35	4.13
Lodging	5.87	2.71	1.12	3.29
Telephone	2.93	1.97	0.55	1.40
Transportation	0.0	0.23	0.34	0.43
Gas & Electricity	2.84	1.97	0.55	2.16
Water	0.18	0.95	0.24	0.78

by multiplying expenditure including tax by one minus the tax rate. These
factors are shown, by category and region, in table 2-2. Any category not
indicated in the table was assumed to be taxed at the "general sales" rate.

Removal of Non-Tax Trade Margins

For all categories except a few such as tobacco which are highly
taxed, the largest share of the difference between retail prices and producer's prices is accounted for by wholesale and retail trade margins.
Data on trade margins were obtained from the 1967 Department of Commerce
367-industry input-output table (BEA, 1974 and 1976). The survey categories were matched to the I-0 industries and weighted average trade margins were computed where necessary using total personal consumption expenditures as the weights.

We updated these trade margins from 1967 to 1973 as follows: for
several broad categories of goods (services have no trade margins by
definition) we compared the change in the consumer price index for each
good from 1967 to 1973 to the change in the wholesale price index for the
same period. The ratio of the change in the consumer price index to the
change in the wholesale price index was assumed to represent the change
in the combined wholesale and retail trade margin. These ratios were
then applied to all the items within each category of goods to update
both the wholesale and the retail margins to 1973. The size of the whole-

Table 2-2. Regional Tax Adjustment Factors, 1972

Category	North East	North Central	West	South
General Sales	0.9419	0.9625	0.9565	0.9583
Food	1.000	0.9858	0.9896	0.9801
Motor Vehicles	0.9419	0.9628	0.9566	0.9704
Admissions	0.9919	0.9888	0.9902	0.9833
Restaurants	0.9360	0.9625	0.9565	0.9587
Lodging	0.9413	0.9729	0.9888	0.9671
Telephone	0.9707	0.9803	0.9945	0.9860
Gas & Electricity	0.9716	0.9803	0.9945	0.9784
Water	0.9982	0.9905	0.9976	0.9922
Transportation	1.000	0.9977	0.9966	0.9957
Gasoline	0.6980	0.7310	0.7380	0.7190
Alcoholic Beverages	0.8800	0.8800	0.8800	0.8800
Tobacco	0.4370	0.5440	0.5900	0.6070

sale margin relative to the retail margin was assumed not to change. The
ratios used are shown in table 2-3.

Matching to Sectors

The final data transformation involved matching the items of the
diary survey and the detailed interview survey to the sectors of the IN-
FORUM model. This was done by referring to the 1967 version of the Standard Industrial Classification Manual for the makeup of the INFORUM sectors, which consist of one or more Standard Industrial Classification
(SIC) sectors. It was then usually straightforward to place the survey
items in the correct INFORUM sector. Although there are a great many
items in the diary data, they map into a small number of sectors, usually
by simple aggregation. Difficulties arose mainly in the interview survey, where it was often necessary to disaggregate survey items into more
than one INFORUM sector. This was sometimes done by obtaining the relative shares of the sectors in 1973 from the INFORUM data base. In some
cases the disaggregation was necessarily arbitrary. Some sectors, it
should be noted, include categories from both the diary and the interview surveys; these cannot be combined for individual observations, however, because no household in the survey participated in both the diary

Table 2-3. Ratios of Changes in CPI to Changes in WPI for Broad
Categories of Goods, 1967 to 1973

Category	(1) WPI 1973 (1967=100)	(2) CPI 1973 (1967=100)	Ratio, (2)/(1)
Food	159.1	141.4	0.889
Other Nondurables	120.5	132.8	1.102
Apparel	125.9	127.1	1.010
Household durables	115.8	118.8	1.026
Other Housefurnishings	125.8	119.0	0.946
Other durables	115.8	121.9	1.053
Autos	119.2	111.1	0.932
Gasoline	151.4	118.1	0.780

and the interview. Aggregation to the sector level is possible in these cases only after separate demand functions are estimated for the two kinds of data.

Other complexities included the necessity to distinguish survey items that properly belong to PCE from items that are really maintenance and repair construction in the INFORUM framework. We also had to be careful to eliminate purchases of used goods from the matching scheme, since the model can take account only of new production.

This sequence of transformations pursued two objectives: to prepare the data for econometric estimation at the level of INFORUM sectors, but to apply all the adjustments and corrections to the survey data at the most disaggregated level possible. Price indexes, tax rates, trade margins and other adjustments were therefore defined at the level of categories in the original data.

The Household Projections

The demographic characteristics we have chosen to characterize households are: the number of members (size); the numbers of adults (18 or older) and children; and the age of the household head. These features define 147 possible different types of households: size and adult/child composition define 21 groups, for each of which there are seven possible age classes. Only 112 types are actually considered, however, because the

sample contains very few households with more than four adults or more
than four children. There are thus four adult size classes (1, 2, 3, 4
or more), five child size classes (0, 1, 2, 3, 4 or more) and seven age
classes; but since all households of six or more members are classed together, not all combinations of adults and children are distinguished. Total
consumption in sector r is estimated by

$$C_{rt} = \Sigma_s \, N_{st} \, C_{rs} \, (Y^*_{st})$$

where Y^*_{st} is the income of a household of type s in year t, C_{rs} is per-
household consumption and N_{st} is the number of such households in year t.
The projection of Y^*_{st} is discussed in the next section; here we consider
the projection of N_{st}.

The Census Bureau projections, series D, E and F, give the complete
age structure of the population in each year but make no assumptions about
the number and type of households. The total number of households is projected by Alterman (1976). It remains then to estimate values of N_{st}
which will be consistent with the projected total numbers of adults, of
children and of households, and with the adult age distribution. Total
population in households is projected on the basis of the share of population in the base year. The non-household population is assumed to consist
entirely of adults. The share of population in households is 98 percent
for all series out to 1980, and thereafter 98.1 percent for F, 98.3 for
E and 98.4 for D (Census Bureau, 1972).

The projection of N_{st} occurs in three stages, corresponding to the
distinction among primary households, childless families and families with
children. A family in the Census definition is a unit of two or more
people related by blood, marriage or adoption. Every one-person household
is then a primary unit, as are units of two or more unrelated people.
This distinction, which is not made in the survey data, is assumed not to
affect consumption behavior; it is used only to simplify the allocation of
children among households.

Primary households are projected to 1990 by the Census Bureau (1975a)
for population series II. We use the age-specific rates of primary households per population derived from these projections and apply them to

series D, E, and F. Beyond 1990, the age-specific rates are assumed not
to change: that is, the increase in primary households is assumed to be
complete, relative to population by age, at that date (the Census projections show an increase of these units out to 1990, but at a decelerating
rate). Two further assumptions are used to account for the population in
these units: (1) they include no children, since children are assumed not
to live alone and are nearly always in households with at least one related member. This assumption may not be exactly true, but the number of
children wrongly assigned will be small; (2) 90 percent of primary units
consist of single individuals (which is consistent with Census estimates)
and the other 10 percent are units of two adults. No larger units are
considered, it being supposed that they are quite infrequent. These assumptions give N_{pt}, the number of primary households in year t; N_{ft} =
$N_t - N_{pt}$, the number of families; and $A_{pt} = 1.1 N_{pt}$, the number of adults
in primary units.

Childless families N_{fot} are projected by assuming the same age-specific rates of childlessness as in the CES data. The number of childless
families will vary in the future as the age distribution of households
changes, but there is assumed to be no further trend toward not having,
or postponing, children. This means that a reduction in children per
adult in the future will be reflected primarily in a different distribution of children among those families having at least one child.

Since anyone 17 years old or less is considered a child, a family is
classified as "childless" even if it includes members 18 years or older
who are the sons or daughters of the household head. There is no attempt
to distinguish these cases or to project the rates at which grown children
leave to form their own households. Individuals are not "followed," that
is, as they age: projections can be made by following people (as Mathematica (1977) has done for one Series II projection), but the procedure
becomes much more complex and does not seem necessary for our purposes,
particularly as there is no longitudinal information on families in the
survey data.

Families with children are all remaining households, N_{fct} =
$N_t - N_{pt} - N_{fot}$. They include all the children, C_t, but how many adults they
include depends on the size distribution of adults among childless fami-

lies as well. Adults have to be assigned to both kinds of families, and children to one kind only, in such a way as to respect the total numbers of adults and children in families and to maintain plausible proportions of adults and children in individual families. This is accomplished by a series of three algorithms, each of which involves one or more time-dependent parameters and is designed to use the minimum number of assumptions, or parameters, required for identification.

The first algorithm uses two parameters to assign children to families. One parameter "discriminates" against large households, making the assignment of another child less likely as there are more children in the family. The algorithm makes no explicit assumptions about preferences for or against one-child families. It achieves its effect primarily by reducing the frequency of many-child households: in fact, the frequency falls to zero for some types of households before the end of the projection period. When this happens, the parameters are modified so as to pass the "pressure" for reduction on to the next-largest families, without, of course, leading to negative frequencies.

A similar procedure could be used to assign adults to families, but it would encounter two problems: first, if applied independently of the assignment of children it might lead to an implausible joint distribution, and second, it could lead to too many one-adult families and too few two-adult families to be consistent with Census projections of husband-wife families. A second algorithm was therefore used to assign adults to families with children: it assigns the joint distribution, and reduces the "pressure" on two-adult households by introducing a time-dependent parameter affecting the relative decline of families with three or more adults. Subject to this time trend, the algorithm respects the proportions of adults and children found in the survey.

Finally, the same time trend is applied in a third algorithm to assign adults to childless families. This procedure is simpler because such families can consist only of two, three, or four or more adults. There is no link in this case to the proportions observed in the survey data.

The distribution of ages of household heads is simultaneously determined in these calculations by repeating the algorithm separately for each age class, with age-specific parameters (the age distribution for primary households is given directly from the age-specific rates used to project the number of such households). Except for the adult/child distinction, the ages of members other than the head are of no concern. No account is taken of possible postponements of births, which would shift the age distribution upward for a given number of children. This is partly because the recent trend to postponement of births is assumed not to continue, and partly because such changes would make it difficult to interpret the age coefficients of the consumption equations, which presumably include some effects due to ages of children.

The details of the calculations and algorithms just described are as follows:

First, population in households (P_h) is calculated from total population (P): P is taken from Census Report #601 for Series I, II, and III, and P_h is calculated from households and average household size projected in report #607. These give the following shares P_g/P:

Series:	I	II	III
1975	0.980	0.981	0.981
1980	0.981	0.980	0.980
1985	0.983	0.983	0.980
1990	0.984	0.981	0.980

Constant shares are assumed beyond 1990.

Primary units (N_p; non-family households) are projected using age-specific estimates N_{pt} from report #607, series II(B) and estimates of population P_t from report #493, for the period 1974-1990. Dividing the II(B) series for N_{pt} by the series E population gives the ratios:

Age	Under 25	25-29	30-34	35-44	45-54	55-64	Over 64
1974	0.0417	0.0805	0.0560	0.0440	0.0720	0.1284	0.2761
1980	0.0537	0.0940	0.0631	0.0507	0.0725	0.1344	0.3041
1985	0.0655	0.1078	0.0696	0.0569	0.0739	0.1393	0.3393
1990	0.0729	0.1251	0.0774	0.0636	0.0752	0.1432	0.3541

Shares for intervening years are interpolated; 1974 ratios are used for 1972-73 and 1990 ratios for 1995-2025. The shares are applied to P_t (by age) for series D, E, and F. N_{pl} (one-person households) is assumed to equal $0.9N_p$, with the remainder being N_{p2}. The total number of adults is $A_{pt} = 1.1 N_{pt}$, with the age distribution indicated.

The number of families is $N_f = N-N_p$, where N is Alterman's projected total number of households. <u>Childless families</u> are projected by $\dfrac{N_{fot}}{N_{ft}} = \dfrac{N_{fo}}{N_f}$ in the base year, all ratios being age specific. Then, $N_{fot} = \Sigma$ (age) $N_{ft}(N_{fo}/N_f)$ for the total number of childless families in year t.

<u>Families with children</u> in year t are $N_{fct} = N_t - N_{pt} - N_{fot}$. Children are assigned to these families as follows: N_{fkt} is the number of families with $k \geq 1$ children.

$$N_{fkt} = \mu_t(1 - k\phi_t)N_{fct}$$

where N_{fct} is total families with children and μ_t, ϕ_t are two time-dependent non-negative parameters. Then N_{fkt}/N_{fct} is smaller, the larger k is, or children are "removed" preferentially from large families. The parameters must respect the constraints:

$$\Sigma_k N_{fkt} = N_{fct},$$ the number of families

and $$\Sigma_k(k \cdot N_{fkt}) = C_t,$$ the number of children

in those families. Solving for ϕ_t, μ_t yields

$$\phi_t = \frac{C_o N_{fct} - C_y N_{fco}}{N_{fct} \cdot \Sigma_k k^2 N_{fko} - C_o C_t}$$

and

$$\mu_t = N_{fct}/(N_{fco} - \phi_t C_o)$$

where o refers to the base period (1972-73 survey). The parameters are also limited by $1 - k\phi_t \geq 0$; since the largest value of k used is four, this means $\phi_t \leq 0.25$. Larger estimates of ϕ_t are replaced by one-fourth and μ_t is recalculated. The parameters are calculated separately for series D, E, and F.

<u>Adults</u> are then assigned to families with children as follows. Let Θ_{mk} be the survey proportion of k-child families ($k \geq 1$) which also have

m adults:

$$\Theta_{mk} = N_{fmko}/N_{fko}$$

where all variables are age-specific. We then assume that there is no change in the relative frequencies of one- and two-adult families, for a given k and age:

$$N_{f1kt}/N_{f2kt} = N_{f1ko}/N_{f2ko} = \Theta_{1k}/\Theta_{2k}$$

while the proportions of three- and four-adult families decline with a time trend exp $(-\delta t)$:

$$\frac{N_{f3kt}}{N_{fkt}} = \Theta_{3k}\, e^{-\delta t}, \quad \frac{N_{f4kt}}{N_{fkt}} = \Theta_{4k}\, e^{-\delta t}$$

This serves the same purpose as the algorithm for children, in discrimi‐ nating against large numbers of adults, but it does not allow one-adult families to increase relative to two-adult families, as would occur if the same sort of algorithm were applied to adults.

For a given value of δ, chosen to yield a smooth decline in large households, each of the terms N_{fmkt} can be calculated from the correspond‐ ing Θ_{mk} and the value of N_{fkt} estimated in the previous step. The value of δ was chosen to guarantee that N_{f2t}, the number of two-adult families, never drops below the number of husband-wife families projected by the Census Bureau, separately for those with and those without children.

The parameter δ is then also used to assign adults to childless fami‐ lies, as follows:

$$N_{f4ot} = e^{-\delta t} N_{f3ot}$$

(four-adult childless families decline relative to those with three adults). Since childless families include at least two adults, this assumption suf‐ fices to calculate N_{f2ot}, N_{f3ot} and N_{f4ot}:

$$N_{f2ot} + (1+e^{-\delta t})N_{f3ot} = N_{fot}, \text{ the number of childless families}$$

$$2N_{f2ot} + (3+4e^{-\delta t})N_{f3ot} = A_{fot} = A_t - 1.1N_{pt} - A_{fct}, \text{ the number of}$$

adults in those families, after subtracting those in primary households and those in families with children. Making all algorithms age-specific simultaneously produces the distribution of age of head of each type of household. For use with the consumption functions, the numbers N_{p1}, N_{p2}, and N_{fmk} are regrouped by total size (merging two-person primary units with two-person families) to yield frequencies N_{mk} ($m \geq 1$, $k \geq 0$) or equivalent frequencies classified by k and total size m + k. In the process, several kinds of large households ($m \geq 3$ and $k \geq 3$ as well as m = 2, $k \geq 4$ and $m \geq 4$, k = 2) are grouped together in the class of six or more members.

Projected Household Numbers by Type and Composition

We first show, in table 2-4, the projected numbers of households by type--the primary units, childless families and families with children. For projecting expenditures, this classification is replaced by a classification of households according to the age of the head, the number of adults and the number of children; expenditure functions are based on total size and the number of children. (The consumption functions do not distinguish families from non-family households.) We next show, in table 2-5, the projected numbers of households, and also the projected mean income for each type of household, using this classification. The age classification is not shown simultaneously, since to include it would make the table seven times as large; instead, the distribution of household heads by age, irrespective of the numbers of adults and of children, is shown in table 2-6.

In all three population scenarios, the number of adults grows appreciably faster than that of children; scenario F in fact shows a steady decline in the number of children, particularly rapid in the last quarter century. Primary or non-family household units increase more rapidly than any other kind of household, but--because these consist entirely of adults, and include a large proportion of the elderly--there is relatively little difference between population projections. The slowest-growing group is that of families including children, which also shows the greatest difference across scenarios. Note that this is not due to assuming that age-specific childlessness will increase when there are fewer children (scenario F): the age-specific rates are held constant for projection, but

Table 2-4. Projections of Adults, Children, and Households by Type
(Thousands), Benchmark Years, Three Scenarios

	Scenario	D	E	F
1975:	Total population	215324	213925	213378
	Adults in households*	141451	140937	142976
	Children (aged 17 or less)	69674	68816	66241
	Total households	71566	71566	71566
	Primary Units	15506	15506	15506
	Childless families	24374	24374	24374
	Families with children	31686	31686	31686
	Average household size	2.95	2.93	2.92
	Average children in family with children	2.20	2.17	2.09
1985:	Total population	243935	235701	230913
	Adults in households	161954	159381	164309
	Children	77834	71722	62101
	Total households	88260	88260	88060
	Primary units	21316	21316	21316
	Childless families	27393	27393	27393
	Families with children	39551	39551	39351
	Average household size	2.72	2.62	2.57
	Average children in family with children	1.98	1.82	1.58
2000:	Total population	285969	264430	250686
	Adults in households	193463	184360	183927
	Children	87930	74914	61871
	Total households	105116	103113	102063
	Primary units	25979	25313	24939
	Childless families	31963	31429	31144
	Families with children	47174	46371	45980
	Average household size	2.68	2.51	2.41
	Average children in family with children	1.86	1.62	1.35
2025:	Total population	367500	303800	264900
	Adults in households	252733	221754	202962
	Children	108887	76122	56772
	Total households	138725	123952	114405
	Primary units	36920	34319	32587
	Childless families	44979	41451	39453
	Families with children	56826	48182	42365
	Average household size	2.61	2.40	2.27
	Average children in family with children	1.92	1.58	1.34

(Continued)

Table 2-4.   (Continued)

Scenario	D	E	F
Average annual growth rate (percent), 1975–2025			
Population	1.069	0.701	0.433
Adults in households	1.161	0.907	0.701
Children	0.893	0.202	-0.309
Households	1.333	1.105	0.943
Primary units	1.735	1.589	1.485
Childless families	1.225	1.062	0.963
Families with children	1.168	0.838	0.581
Average household size	-0.245	-0.399	-0.504
Average children in family with children	-0.272	-0.635	0.889

*Excluding adults in institutions and armed forces. All children are assigned to households.

population F is slower growing and older on average, which raises the over-all rate of childlessness. In all three scenarios, every type of household increases more rapidly than the total population, because average household size falls, from about 2.9 persons to only 2.6 (scenario D) or only 2.3 (scenario F). The number of children per household including children falls below 2.0 by 1985, and continues falling in the two lower-growth pro-jections: for scenario D, however, there is a slight expansion in children per family during 2000–2025, when there is a 24 percent increase in the number of children. By the year 2025, there are almost twice as many chil-dren under assumption D as under F, but there is only a 25 percent differ-ence in the number of adults, and only a 21 percent difference in the num-ber of households.

The algorithms for assigning children and adults to households ensure that in all scenarios, there are no households of three or more adults and four or more children, or vice versa. In scenario F, where the number of children is actually projected to decline, there are from the year 2000 onward no households of any size with four or more children. In general, the larger the family size, the more difference there is among scenarios in any year. Although one-person households increase more rapidly than any other kind in every scenario, out to the year 2025 there are still

Table 2-5. Household and Income Projections: Numbers (thousands) and Incomes (1971 dollars) of Households Classified by Number of Adults and Children, by Scenario and Year

1975

Scenario:	D				E				F			
No. of adults:	1	2	3	4+	1	2	3	4+	1	2	3	4+
No. of children 0	13955	20698	2868	2360	13955	21052	2674	2200	13955	19647	3444	2834
	4092	7139	8824	11965	4074	7106	8784	11911	4032	7032	8693	11788
1	1380	7096	1580	1107	1380	7096	1580	1107	1380	7096	1580	1107
	4917	8267	11203	13041	4895	8230	11152	12982	4844	8144	11036	12847
2	1002	7606	1063	959	1002	7606	1063	959	1002	7606	1063	959
	5089	9381	10949	12550	5066	9339	10899	12493	5013	9242	10786	12364
3	484	4209	1246	0	484	4209	1246	0	484	4209	1246	0
	5451	9791	11530	–	5426	9747	11478	–	5370	9646	11359	–
4+	471	3484	0	0	471	3484	0	0	471	3484	0	0
	5234	9540	–	–	5210	9497	–	–	5156	9398	–	–

1985

Scenario:	D				E				F			
No. of adults:	1	2	3	4+	1	2	3	4+	1	2	3	4+
No. of children 0	19185	26283	2267	974	19185	28444	756	325	19185	24942	3206	1377
	5042	8707	10721	14497	5137	8872	10923	14771	5241	9050	11145	15071
1	2175	11307	1139	798	2494	12967	1307	916	2997	15581	1570	1100
	6052	10145	13657	15898	6166	10336	13914	16197	6292	10546	14197	16526
2	1380	10615	695	627	1398	10749	704	635	1425	10961	713	648
	6246	11470	13347	15299	6364	11686	13598	15587	6493	11924	13874	15904
3	594	5352	731	0	503	4538	620	0	361	3255	445	0
	6697	11955	14056	–	6323	12180	14321	–	6962	12428	14611	–
4+	469	3469	0	0	300	2221	0	0	35	257	0	0
	6381	11629	–	–	6501	11849	–	–	6633	12089	–	–

Table 2-5 (continued)

2000

Scenario:	D				E				F			
No. of adults:	1	2	3	4+	1	2	3	4+	1	2	3	4+
No. of children
0	23381	25750	7581	1228	22781	30230	3211	520	22445	28771	4188	679
	7260	12518	15534	21071	7396	12779	15886	21501	7490	12950	16122	21786
1	3131	16376	571	400	3730	19509	680	476	3879	20291	707	496
	8694	14547	19513	22715	8871	14844	19911	23179	8989	15040	20175	23486
2	1772	13738	317	286	1777	13775	318	287	1772	13733	317	286
	8961	16417	19070	21860	9144	16752	19460	22306	9265	16974	19718	22602
3	668	6149	294	0	477	4388	210	0	423	3891	186	0
	9613	17097	20083	-	9809	17446	20493	-	9939	17677	20765	-
4+	414	3061	0	0	89	657	0	0	0	0	0	0
	9117	16617	-	-	9303	16956	-	0	-	-	-	-

2025

Scenario:	D				E				F			
No. of adults:	1	2	3	4+	1	2	3	4+	1	2	3	4+
No. of children
0	33228	33841	14373	459	30997	34484	10077	322	29328	34213	8236	263
	9959	17440	21603	29372	9959	17660	21962	29861	10108	18110	22578	30699
1	3739	19609	129	91	4123	21620	143	100	3709	19451	128	90
	12421	20768	27817	32381	12664	21175	28362	33015	13060	21836	29248	34047
2	2186	17001	75	68	1904	14808	65	59	1678	13055	58	52
	12797	23421	27186	31162	13048	23880	27718	31772	13456	24626	28584	32766
3	873	8112	74	0	481	4465	41	0	400	3713	34	0
	13732	24382	28630	-	14001	24860	29190	-	14439	25637	30103	-
4+	580	4293	0	0	45	332	0	0	0	0	0	0
	12996	23688	-	-	13251	24152	-	-	-	-	-	-

Table 2-6. Projected Age Distribution of Household Heads (thousands) by
Scenario and Year

Age	1975			1985		
	D	E	F	D	E	F
Under 25	6015	6029	5974	7778	8118	8483
25-29	7773	7784	7738	10483	10762	11029
30-34	6510	6514	6499	8432	8193	7757
35-44	12123	12121	12128	15132	14291	12995
45-54	13118	13112	13138	14731	15024	15563
55-64	11559	11548	11589	13244	13393	13768
Over 64	14470	14459	14503	18263	18281	18467

Age	2000			2025		
	D	E	F	D	E	F
Under 25	9982	9639	9199	11780	9957	8280
25-29	11769	11794	11712	14461	12650	11053
30-34	9425	8831	8636	11848	9303	8116
35-44	18102	16295	15763	21985	16427	14153
45-54	20509	20890	20968	22738	20969	18958
55-64	14599	14869	14950	21869	21003	20493
Over 64	20733	20798	20834	34048	33646	33356

roughly twice as many two-adult as one-adult households. By the end of
the projection period, there are more households without children under 18
than with children; among those with children, families with only one child
are most numerous. Two-child families are nearly as common in scenario D
but much less frequent in scenario F.

The age distribution of household heads, shown in table 2-6, does not
change through time, or differ between scenarios, nearly so much as the
age distribution of the entire population: much of the difference between
one demographic path and another is in the number of children. Moreover,

the aging of the household-head population is partly offset by the increase in young, one-person households. Thus, the only notable overall change is an increase in the median age of heads, which in all scenarios rises from about age 40 up through the year 2000, to 50 or more by the year 2025. The aging of the population has the greatest effect on demand in those sectors which do not depend much on how people are organized into households, namely education and health care: changes in the age distribution of household heads have their greatest effect on the demand for housing and durables.

Income Projections

Ridker and Watson's time-series estimates of future consumption in each sector are based on projections of per capita disposable income, which in turn depend on projected labor force and average labor productivity. Only mean income per person is projected; the distribution of income does not matter. Our cross-section equations, in contrast, require an estimate of income for each of the different types of households distinguished by the demographic characteristics, so that total consumption in a sector is sensitive to the distribution of income among households, because the equations are nonlinear in income, and also because even the linear income terms differ among size classes.

Income for each type of household is projected by first converting per capita income into average income per household, using the number of households, and then using two assumptions. The first is that, in the absence of changes in the composition of the population of households, each type of household would maintain its relative position in the distribution, or the same ratio of its income to mean household income. That is, there are assumed to be no technological or institutional factors affecting the shape of the distribution or the position within it of different kinds of households. The income distribution changes only in response to demographic changes, including changes in the age distribution which may be associated with differences in productivity. (Since education and other human-capital variables are not included in the consumption equations, it is not

necessary to guess how the relative income associated with a given schooling level will change as the population becomes more educated.)

The second assumption is needed to take account of the fact that if
the relative numbers of different kinds of households change, their relative incomes must also change in order to leave mean income per household
unchanged. Whereas the first assumption assigns to each type of household
a factor relating its income to the mean--defined in the cross-section for
the base year--the second defines a factor for each future year applic-
able to all households. The assumptions are expressed by the equations:

$$Y^*_{so}/\bar{Y}^*_o = \alpha_s$$

$$\Sigma_s N_{st}/\Sigma_s \alpha_s N_{st} = \gamma_t$$

$$Y^*_{st} = \alpha_s \gamma_t \bar{Y}^*_t$$

where t refers to the year (t=0 corresponds to the survey period, 1972-73)
and s to the type of household. N_{st} is the number of households of type
s in year t, described in the previous section; Y^*_{st} is the income of a
type -s household in year t. (all households of that type being assumed
to have the same income), and $\bar{Y}^*_t$ is mean household income in year t. Then
α_s is the base-year relative position in the income distribution, and γ_t
is the year-specific adjustment to take account of changing frequencies
N_{st}.

This procedure takes no account of income variation among households
of a given type, but it does account for two of the principal sources of
income differences--household size and age. The assumptions are consistent
with Stoikov's (1975) observation that a large share of income variation
is due to age differences, and with Paglin's (1975) observation that changes in the shape of the income distribution are associated with changes
in the age distribution, even if everyone's lifetime income profile is
unchanged. The overall distribution becomes slightly more or less equal
in response to demographic factors but does not change appreciably in
shape.

One further adjustment to the income estimates is required for projection of sectoral consumption. If the estimate of income for each

household type is used to project consumption per household, and total consumption is estimated by multiplying by N_{st} and summing across household types, the sum across sectors—the estimate of total expenditure—will not necessarily match the total estimated from time-series equations for the same scenario and year, which are considered correct for the given income level.

In order to match the totals, the estimate of per capita disposable income is raised or lowered and the distribution of household incomes is recalculated. This procedure continues until the target and calculated total PCE estimates are equal to within one percent. The effect of this adjustment is to base the sectoral consumption projections on a "projected" per capita income which is not necessarily consistent with the projections of productivity and labor force: per capita income becomes an instrumental variable rather than a forecast. However, as Table 2-7 shows, the income value reached by iteration is usually quite close to the target value projected. At the end of the period the difference is less than $100 per capita for scenario F, just over $200 for E, and only $100 for D. (The cross-section estimates are less dispersed, making it appear that demographic changes have less effect on per capita income, but in fact the same productivity assumptions are used in both projections.) Table 2-5, above, shows an example of the final household income distribution. From 1985 on, household incomes are, of course, systematically slightly higher as population growth is lower. The shape of the distribution is, however, very little affected. The ratio of income in the year 2000 to income in 1975, for a given type of household, varies only from 1.742 to 1.768 in scenario D, and from 1.828 to 1.858 in scenario F. A comparison by age groups shows only slightly more variation.

Income-Demographic Relations

The projection procedures we have developed allow for only two forms of interaction between income and demographic factors. The first is that the projections begin with assumptions about average productivity per worker; income per capita then depends on the labor force participation rate, which in turn is inversely related to the rate of population growth

Table 2-7. Projected and Calculated Per Capita Disposable Income by
Scenario and Year

Scenario: Year	D		E		F	
	Projected	Calculated	Projected	Calculated	Projected	Calculated
1972	2779	2912	2779	2912	2779	2912
1975	2605	2682	2605	2682	2605	2682
1980	3102	3108	3194	3133	3280	3236
1985	3299	3305	3401	3325	3619	3444
1990	3623	3625	3883	3905	4090	3948
1995	4205	4165	4490	4398	4747	4593
2000	4788	4706	5096	5041	5403	5269
2005	5239	5180	5596	5459	5974	5766
2010	5691	5655	6095	6050	6545	6262
2015	6041	5939	6501	6399	7048	6625
2020	6392	6419	6907	6748	7552	7173
2025	6794	6910	7300	7086	7852	7473

because it is inversely related to the proportion of children in the population. The second interaction connects the demographic structure of the
population to the distribution of income among households, once the average level is determined. Both effects run from demographic factors to
income. No consideration is given to effects running the opposite way:
that is, to effects of income on the demographic structure or rate of
growth.

At least two such reverse effects are likely. One is that higher
incomes lead families to have more children, at least when their relative
position in the income distribution is unchanged. (A rise in position
might have the opposite effect on an initially poor family.) Thus faster
income growth should lead to faster demographic growth. The other effect
is that higher incomes should lead to the creation of more households for
a given number of adults, as young (or possibly old) adults can afford to
form or maintain separate households rather than living with their parents
(or children). Higher income growth would thus reduce the average number
of adults per household; although if it also increased the number of children per household, its effect on average total size would be ambiguous.

These effects are not taken into account in our projections because
the Census Bureau's population projections are not explicitly income-sensi-

tive (although their underlying fertility assumptions may correspond to different expectations about income). It is outside the scope of this study to make new population projections based on different income projections; it would be extremely difficult, and quite arbitrary with the information available, to project the structure of a given population as a function of income. And the expenditure data do not include any longitudinal information on families; we would simply have to project different adult-size household distributions to reflect different rates of household formation in response to income changes.

This limitation probably does not seriously affect the projections based on continued growth of productivity, since scenarios D, E and F differ considerably and all are, for the purposes of Census Bureau projections, compatible with continued income growth at rates comparable to those of the recent past. The no-income-growth scenarios, however, are based on the same assumed population structures, when in fact a cessation of income growth might lead to a different distribution of households. In the absence of income-sensitive demographic projections, or of reasonably clear relations on which to base them, we have chosen the simpler course of ignoring possible effects of this type.

The Consumption Functions: Specification

The household data were used to estimate consumption (expenditure) functions for 117 sectors. Projections were then made by combining these functions with projections of numbers of households and of income. Three choices had to be made in estimating the equations and deciding which ones to retain: these concern the level of aggregation, the functional form and the reasonableness of the projections.

Level of Aggregation

Equations could have been estimated for variables defined in the household data at a highly disaggregated level, and the estimates then combined to match the INFORUM sectors. This approach would permit, in effect, a great deal of flexibility in functional form, since the coefficients of the sectoral (aggregated) equations would vary as the component variables

changed in relative importance. Apart from the difficulty of deciding
which component variables to use, however, this approach has the disadvantage that it would be difficult to judge the reasonableness of the aggregate equation or the significance of the explanatory variables. We chose,
therefore, to aggregate the data--to create the sectors--first, and then
to estimate one equation for each sector.

There is one exception to this procedure, because the CES diary data
do not include total household expenditure, whereas the interview data do
not include details of food expenditure and other frequent household purchases. It is therefore not possible to relate the diary sectors directly
to total spending and income. The solution adopted is to use the interview data to estimate functions for two aggregates--total expenditure on
food eaten at home, and total expenditure on food away from home--and then
to relate the individual diary sectors to these two aggregates. This makes
the equations for the diary sectors somewhat more complicated than those
for the interview sectors, which are related directly to an estimate of
permanent income.

Functional Specification

If a set of expenditure equations is to respect the budget constraint,
they must all be of the same form and must include a term linear in income
(or a related variable). If higher powers of income are included, the
equations will eventually lead to absurd results if used to project over
a period of substantial income growth: this occurs because the sum of
the coefficients must be zero, so at high enough income levels projected
demand in a sector must either become negative or exceed total income.
Lower powers of income can be used, however, because their influence tends
to zero as income rises; their inclusion allows the functions to be curvilinear at low incomes, although at high enough incomes the marginal budget
shares become constant.

So far as _income_ ($Y*$) is concerned, the equations estimated have the
form

$$\beta_o + \beta_1 Y* + \beta_2 (\log Y*)^{-1}$$

where the log-inverse term introduces curvilinearity which dies away more

slowly than a term in $1/Y*$. This specification allows for saturation as
a share of the budget (if $\beta_1 > 0$) or in absolute amount (if $\beta_1 = 0$ but
$\beta_0 > 0$). Saturation is approached from below if $\beta_2 < 0$ and from above if
$\beta_2 > 0$; any combination of signs of the coefficients is allowed except
that β_1 cannot be negative. Since this specification captured the principal effects of interest, no alternative forms were tested.

The constant and linear terms alone are consistent with demand functions derived from the Stone-Geary utility function (Lluch, Powell and
Williams, 1977) in which consumption of each sector must exceed a minimum
level which, if positive, is sometimes interpreted as a subsistence or
essential expenditure. With the introduction of the log-inverse term, the
equations no longer correspond exactly to a utility function, but this is
not considered important; they may approximate a utility function closely.
The cross-section data do not include information on prices, and no attempt is made to estimate price responses, so a complete utility specification is impossible.

For the interview sectors, the variable $Y*$ is an estimate of permanent
income, formed by taking [var (C).Y + var (Y).C]/[var (C) + var (Y)] where
C and Y are respectively total household expenditure and total income, and
var (C) and var (Y) are their variances in the sample. The object of this
linear combination of income and expenditure is to remove some of the
transitory income variation which would otherwise bias β_1 downward; C gets
a larger weight than Y because it varies less in the sample and therefore
presumably includes less transitory noise. This is an approximation to a
more elaborate estimator (Musgrove, 1979), which is based on instrumental
variables, C and Y being first regressed against a set of explanatory
variables.

For the diary sectors, the equation includes food at home and food
away from home and their respective log-inverse terms. These aggregates
in turn are functions of $Y*$ and $(\log Y*)^{-1}$. This makes the diary sectors
functions of $Y*$ whose coefficients change as the share of income devoted
to food (and the balance between food at home and food away) vary.

Demographic effects could be introduced into the consumption equations in many different ways, with or without interaction with the income

terms. We distinguish three demographic descriptions of the household: its size (number of members), its composition (number of children for a given size) and its age (measured by the age of the household head).

Household size is expected to be the most important of these characteristics, since together with income it determines income per person, and since there may be significant economies or diseconomies of scale in consumption for some sectors. To avoid searching for the best way to combine income and size in a single equation for all households, we chose to estimate the consumption functions separately for each of six household size classes (all households of six or more members constitute the sixth class, since households of more than six members are infrequent in the sample). The coefficients β_o, β_1 and β_2 differ by size, so that the "income effect" for a given sector depends on the composition of the population. In this, we follow the practice of Houthakker and Taylor (1970), who also estimated separate equations for each household size class and found the coefficients generally to differ significantly. Separate estimation by size class removes the distinction between income effects and scale effects, in each equation.

Age is included in the consumption function as a set of six dummy variables distinguishing seven age classes (under 25, 25-29, 30-34, 35-44, 45-54, 55-64 and over 64). The use of dummy variables allows for non-linear and even non-monotonic effects. The 35-44 group is used as the comparison or base class.

Household composition effects are also introduced by dummy variables, which indicate the number of children in the household. The number of variables included varies with the size of the household, so that the maximum number of children allowed is one less than the total number of members (every household is assumed to include at least one adult). "Children" are members aged 17 or younger, members aged 18 or over being considered adults. Experiments were made with a three-way age classification of children corresponding to pre-school, primary school and secondary school-age children, but when so many variables are included in the equation, very few are significant. The adult/child distinction appears to capture the principal effects of composition.

Since the dependent variable in the consumption function is expenditure in a sector, the coefficients of the dummy variables are absolute (dollar) differences between age classes or between households of different compositions. Because these differences are independent of income, they become smaller relative to expenditure as income rises. This means that age and household composition effects tend to "die out" as income rises, so their effect may be understated in long-term projections. However, income only slightly more than doubles over the projection interval, while this form is easier to estimate than one maintaining constant proportional effects, so the constant absolute difference specification is retained. Differences among household size classes, in contrast, are magnified as income rises.

The equations were estimated with an intercept term which could differ among the four regions of the United States identified in the data (Northeast, North-Central, South and West). Since the population projections did not include regional differences in growth rates (due either to fertility or mortality differences or to interregional migration), these coefficients were not analyzed further. They were included in the specification to avoid bias in the income and demographic coefficients which might result from regional differences in those variables, associated with differences in the dependent variable.

The expenditure functions took the following form when estimated from the interview data:

$$C_{rs} = \beta_{ors} + \sum_{m=2}^{4} \beta_{orsm} G_m + \beta_{1rs} Y^* + \beta_{2rs} (\log Y^*)^{-1}$$

$$+ \sum_{\substack{i=1 \\ i \neq 4}}^{7} \lambda_{rsi} L_i + \sum_{\substack{j=1 \\ j \leq N-1}}^{4} \kappa_{rsj} H_j + U_{rs}$$

When they were estimated from the diary data, they took the form:

$$C_{rs} = \beta_{ors} + \sum_{m=2}^{4} \beta_{orsm} G_m + \beta_{1hrs} F_h + \beta_{2hrs} (\log F_h)^{-1}$$

$$+ \beta_{1ars} F_a + \beta_{2ars} (\log F_a)^{-1} + \sum_{\substack{i=1 \\ i \neq 4}}^{7} \lambda_{rsi} L_i + \sum_{\substack{j=1 \\ j \leq N-1}}^{4} k_{rsj} H_j + U_{rs}$$

The subscript s refers to household size (s = 1, 2, 3, 4, 5, 6 or more), and r indicates the sector number; the subscript indicating the individual observation or household is suppressed. The variables are:

C annual household expenditure, in 1971 dollars

G_m dummy indicating region, m = 2(North Central), 3(South), or 4(West). The base class is m = 1(Northeast)

Y^* estimate of permanent income, calculated as

$$\frac{C \cdot var(Y) + Y \cdot var(C)}{var(Y) + var(C)}$$

    where C = total expenditure and

        Y = total income, both in 1971 dollars per year

F_h total annual expenditure on food consumed at home, in 1971 dollars

F_a total annual expenditure on food consumed away from home, in 1971 dollars

L_i dummy indicating the age of the household head, i = 1(under 25), 2(25-29), 3(30-34), 5(45-54), 6(55-64) or 7(over 64) The base class is i = 4(35-44)

H_j dummy indicating the number of children (aged 17 or less) in the household, j = 1(one child), 2(two children), 3(three children), or 4(four or more children). The condition $j \leq N-1$ assures that a variable is not included in the equation if it would always be zero (because every household must include at least one adult)

U stochastic residual

Income and Demographic Effects in the Equations

Four effects can be distinguished in the expenditure equations. Since the equations are estimated separately for each of six size classes, each of three effects--those due to income, age and family composition--is size-specific, whereas the effect of a change in size depends on the coefficients of the other three effects.

The income effect in size class s is:

$$dC = [\beta_{1s} - \beta_{2s} (Y*\log^2 Y*)^{-1}] \, dY*$$

for an interview sector, and for a diary sector.

$$dC = [\beta_{1hs} - \beta_{2hs} (F_h \log^2 F_h)^{-1}] \, dF_h$$

$$+ [\beta_{1as} - \beta_{2as} (F_a \log^2 F_a)^{-1}] \, dF_a$$

where F_h and F_a are respectively food at home and food away from home; dF_h and dF_a are in turn functions of $dY*$, of the same form as dC.

The age effect, on moving from age class i to class j, is

$$\Delta C = \lambda_{js} - \lambda_{is}$$

where the λ are coefficients of the age dummy variables. Similarly, the effect of a change in composition (varying the numbers of adults and children while keeping total size s constant) is

$$\Delta C = \kappa_{js} - \kappa_{is}$$

where the κ are coefficients of the dummy variables for different numbers of children. (No composition effect is possible, of course, for one-person households).

The effect of a change in household size, holding constant income, age, and the number of children, is then

$$\Delta C = (\beta_{1s} - \beta_{1s'}) \, Y* + (\beta_{2s} - \beta_{2s'}) \, (\log Y*)^{-1}$$

$$+ (\lambda_{js} - \lambda_{js'}) + (\kappa_{js} - \kappa_{js'}) + (\beta_{os} - \beta_{os'})$$

where s and s' are the two size classes compared, and j is the index for age and composition classes. This expression corresponds to a change in the number of adults in the household only; if the number of children also changes, the term κ is replaced by $(\kappa_{js} - \kappa_{is'})$, where the j-class of children corresponds to size s and the i-class to size s'. For diary sectors, a size change includes two terms in differences in β_1 and two in differences in β_2. A significant "size effect" can arise from differences among size classes of any of the three sets of parameters β, λ, and κ (there can also, of course, be differences in the constant term β_o of the equation). This interaction makes it difficult to separate income and demographic effects in the individual equations. Projections, however, were made once with, and once without, anticipated productivity growth, as described in chapter 1. There is then no "income effect," but only a "demographic effect" in the no-growth projection; although this does not mean that income is exactly constant, because income itself is influenced, through participation rates, by demographic change.

This specification of the consumption function was tested in two ways: by using observed consumption (C) or income (Y) in place of the permanent income estimator Y*, and by varying the specification of the household composition. The first test showed C or Y* systematically to give a better fit to the sample (higher R^2) than Y, while Y* performs almost as well as C and is to be preferred for projection. The second test showed composition to affect expenditure significantly for a number of sectors and size classes, a simple adult/child distinction sufficing to show the effect.

The use of a single cross-section to estimate the equations forced us to ignore Lester Taylor's (1971) dictum that "any consumption function worth its salt these days is dynamic." The data contains no information on past incomes or income changes, so--in contrast to time-series equations estimated from aggregate data--only current income effects can be taken into account. The data contain some information on wealth, but it is too sparse to be incorporated into a model of adjustment to income changes or to serve as a proxy for "state" variables as in the Houthakker-Taylor analysis of time series. No distinction is made between habitual purchases and those which are impulsive responses to income changes.

Estimation of the Equations

Because the households participating in the survey were selected independently, the residuals U_{rs} can be assumed to be uncorrelated across observations. They are also assumed to be uncorrelated across equations, because most sectors take only small shares of a household's total expenditure, and because not all sectors were estimated from the CES data. (The residual in any one equation therefore need not be correlated even with the sum of the residuals from other equations.) Ordinary least-squares (OLS) regression could therefore be expected to give unbiased estimates of the parameters, provided the expenditure function is correctly specified. OLS estimation in these circumstances might still be inefficient, because the residuals U_{rs} might not have constant variance, but might increase in size with the income variable (s). We did not test the regressions for heteroskedasticity, and therefore cannot be sure that the standard errors of the parameter estimates are as small as possible. (If these or similar equations are used in further research, it might be desirable to divide them through by Y^* (or F_h) to normalize on the income variable. Of course, if the specification is incorrect, this normalization would also change one or more of the parameter estimates.)

The dummy variables L_i, and the variables H_j, cannot be significantly correlated within each group. They could be correlated across groups (particular numbers of children being associated with particular ages of the head), but those correlations were in fact never high enough to present a serious problem of multicollinearity. Since the equations were estimated separately for each household size class, there is no problem of correlation between size and age, or between size and the number of children. The income variables also were not highly correlated with the age or composition dummy variables, and while there is some correlation between an income variable and its log-inverse, it was never dangerously high. More surprisingly, perhaps, F_h and F_a were never very highly correlated, in part because they are substitutes at any given income level as well as complements in that they both grow with income. Due to individual tastes, there is also much variation in the proportion of food consumed at home, even given income, age, family size and family composition.

Use of the Equations for Projection

The regression equations estimate mean household expenditure by sector, for given types of households--that is, conditional on the values of the independent variables Y^* (or F_h and F_a), G, L and H. Inter-household variance in expenditure within a household type is then equal to the residual variance, var(U). Since we are interested only in projecting total expenditures, we ignored this variance; total expenditure among households of a given type is simply the product of mean expenditure and the number of households of that type. Total expenditure among all households is then estimated as the sum of those products over all household types, or

$$C_r = \sum_s \sum_m \sum_i \sum_j N_{smij} \cdot C_{rs}(Y^*, G_m, L_i, H_j)$$

where N_{smij} is the number of households of size s with particular values of G_m, L_i and H_j, and where income Y^* is in turn a function of the mean household income $\overline{Y}^*$ and the entire distribution of the N_{smij}. When r is a sector estimated from the diary data, Y^* is replaced by $F_h(Y^*, G_m, L_i, H_j)$ and $F_a(Y^*, G_m, L_i, H_j)$. Although the income variables are in principle continuous, only a finite number of values were projected, one for each household type. Total expenditure C_r can therefore be found by a discrete sum rather than by an integral over Y^*.

Three other points need to be emphasized here. The first is that, for some sectors, part of the information came from the interview data and part from the diary data, neither one accounting for all the components of the sector. In those cases, one equation was estimated for each source, and the projected values were summed. Second, for the diary sectors, values of F_h and F_a were projected for each household type, and then these projected values were used to project expenditure by sector. This means that F_h and F_a were treated like two additional sectors, to be estimated in a first stage from the interview data. Third, since regional projections were not made, the equations were projected using an intercept term which maintained the base-year distribution of households among regions. That is,

$$\beta_{ors} + \beta_{ors2} \overline{G}_2 + \beta_{ors3} \overline{G}_3 + \beta_{ors4} \overline{G}_4$$

is constant for each sector, $\overline{G}_2$ being the share of households in region 2, and so on. This means that the sum over m is suppressed; N_{smij} becomes just N_{sij}, and C_{rs} becomes a function of the income variables, L_i and H_j only.

Goodness of Fit

The CES data were originally used to fit expenditure functions for 117 sectors. Of these, only 76 were retained, some being discarded because the equations explained very little, and others because of estimated coefficients of the wrong sign, exclusion of important relative price effects, or other reasons. The fit was judged satisfactory for the 76 sectors finally estimated from the CES, although the R^2 statistics were sometimes very low. Separate estimation by size class removed much of the inter-household variation, and thereby probably improved the fit substantially. Table 2-8 summarizes the R^2 statistics by sector and household size class. The F-statistics were almost invariably statistically significant, and the number of observations was always large enough that no adjustment is required for degrees of freedom, as is evident from this tabulation of sample size by household size and data source:

Data	Size:	One	Two	Three	Four	Five	Six or More
Interview		4575	5511	3192	2896	1728	1731
Diary		4975	6314	3634	3327	2033	1898

In general, table 2-8 shows slightly better results using the diary data than using the interview data; in particular, there are only six regressions for which less than one percent of the variance is explained, versus 51 cases using the interview data. This is, we expect, partly because there are four rather than only two "income" variables in the diary regressions, and partly because the diary sectors are typically more important and are bought by all or nearly all households. The regressions explaining the diary income variables, food at home and food away from home, usually yield R^2 statistics of 20 to 30 percent; the higher income elasticity of food away from home is probably what makes it easier to explain:

Table 2-8. Summary of R^2 Statistics (Shares of Variance Explained)
of Expenditure Equations, by Source, Sector and Household
Size

| Range of R^2 | Number of Size Classes | Sectors, from: | |
		Diary Data	Interview Data
over 0.25	all six	7, 23, 24, 25, and 27	39 and 166
	four	none	76
0.05-0.25	all six	29, 30, 31, 32, 33, 49 and 67	38, 45, 53, 72, 77, 126, 133, 145, 147 and 155
	five	26, 170	40, 52, 54, 73, 123, 158, 162 and 170
	four	none	59, 125, 169, 174 and 176
	three	none	49 and 168
	two	none	35
	one	28, 34 and 74	39, 80, 124, 146 and 165
0.01-0.05	all six	56, 57, 58, 61, 73, 82, 87 and 124	74, 116 and 130
	five	28 and 34	80, 97, 139, 153 and 165
	four	none	103, 124, 146 and 152
	three	37 and 40	128, 148, 151, 154 and 168
	two	none	35, 49, 122, 125, 131 and 140
	one	26	43, 54, 59, 73, 123, 142, 162 and 169
below 0.01	all six	none	37 and 157
	five	none	43 and 142
	four	none	81, 122, 131 and 140
	three	37 and 40	128 and 151
	two	none	103 and 154
	one	none	49, 139 and 153

Household Size:	One	Two	Three	Four	Five	Six or more
F_h (home)	0.1960	0.2213	0.2223	0.2386	0.2304	0.2178
F_a (away)	0.2969	0.3090	0.2794	0.3033	0.3062	0.2484

There does not seem to be much systematic association between household size and the goodness of fit of an expenditure equation. Roughly the same size R^2 statistic is usually obtained in all size classes, for a given sector. (The arbitrary intervals for the statistic in the table partially hide this result.) There is a strong association between goodness of fit and the importance of a sector to consumers: those best explained are clothing and shoes and the major categories of food at home. Rather surprisingly, it was possible to explain about ten percent of the variance for some important durables, such as furniture, automobiles, tires and tubes, and some household appliances. The sectors for which the equations explain almost nothing are typically bought by very few households; most sample observations are zero, but since the equation is fitted to all observations, it predicts some expenditure for all households. In these circumstances, mean and total expenditures may be fairly accurately predicted even if the variation among households is not explained.

Reasonableness of the Projections

Even when R^2 was satisfactory for a sectoral equation, β_1 could be negative (even if not distinct from zero) and negative values for some of the demographic coefficients could lead to negative estimated expenditure for some types of households. (When R^2 is low and β_1 close to zero, β_0 and β_2 may change sign between one household size class and another, leading to implausible projections.) These equations cannot be used for projections, so we drew on the time-series equations of Ridker and Watson, which exclude composition effects but give a better picture of income effects. These aggregate functions were also used for all energy sectors (gasoline, natural gas, heating oil, coal and electricity) even though the CES data yielded good fits and reasonable coefficients. This decision was based on the fact that the Ridker-Watson equations incorporate the anticipated effects of the increase in energy prices that began late in 1973. The cross-section data include only about eight months of observations following the oil price rise, far too little to observe any long-term adjustment in consumption, and they take no account of anticipated technological changes.

The 76 sectors which yielded consumption functions considered satisfactory for projection accounted for 47.5 percent of PCE in 1972, the survey base year. For another 51 sectors, accounting for 18.5 percent of PCE, aggregate projections were used (these include 37 sectors for which CES equations could be estimated and 14 for which the household data contain no information).

Up to this point, the decision to retain or replace a consumption function was based on a priori considerations: sign of the income term, size and significance of the demographic coefficients, goodness of fit, and the presence of negative estimates for some cells of the population. The reasonableness of the functions was, of course, also judged after completing the projections for the entire population. The Ridker and Watson time-series equations served as the first point of comparison. The projections were also compared to exogenous estimates for sectors—particularly food—for which physical saturation can be expected. Where simple projection with the cross-section equations would lead to unreasonably high values (because saturation is not evident in the sample but must occur at income levels foreseen during the next half-century), the equations were modified by suppressing or changing individual coefficients, by adding time trends, or in some cases by forcing an overall fit to the projections obtained by Ridker and Watson. These adjustments override the income effects estimated from the sample while preserving the demographic effects so far as possible. In a few cases, the income coefficient β_1 was modified. More commonly, the entire estimating equation was multiplied by a function of the form exp $(f(t))$, where $f(0)=0$ so that no adjustment occurs in the base year. The function $f(t)$ was chosen to match exogenous projections; these are discussed in chapter 6. All such adjustments were estimated for scenario E and then applied to scenarios D and F, so as to keep the same set of assumptions in comparing the three scenarios.

In total, consumption functions were estimated for 45 sectors defined in the diary data, and for 99 defined in the interview data. Since 27 sectors appear at least partly in both sources, there are 117 different sectors for which equations were estimated. No estimates were made, of course, for sectors in which households do not buy anything directly (178-183, 185, 187-188, 190-195).

The list of 117 sectors was reduced, for projection, for the different reasons indicated above:

Energy sectors (14, 69, 70, 160, 161) were estimated using Ridker and Watson's equations, since these take price rises into account and anticipate reduced physical per capita demand.

Publicly provided sectors include education (176 and 189) and health (66, 143, 144 and 175) for which new demographically-sensitive equations were estimated using other data instead of, or in addition to, that provided by the CES. Households account for only a small share of expenditure in these sectors. The same procedure was used for housing construction (including investment sectors as well as part of sector 140, mobile homes). These projections are described in chapters 3, 4, and 5.

Negative projections of total expenditure led us to drop 17 sectors (16, 22, 46, 48, 51, 62, 68, 78, 83, 90, 94, 96, 99, 117, 119, 120 and 137) and use time-series equations instead. These are sectors for which the R^2 statistic is very low (often below 0.05), the income coefficient β_1 is zero or negative, and the mean expenditure per household in the survey is often less than one dollar per year.

Negative estimates for 35 or more of the 112 household types led to use of the Ridker-Watson equations for seven sectors (10, 60, 93, 98, 101, 102, and 172). None of these is sensitive to demographic factors, and mean household expenditures are usually below five dollars per year.

Large discrepancies between our projection and that given by Ridker and Watson, together with an absence of significant demographic effects, led us to borrow their equations for another seven sectors (21, 106, 115, 127, 129, 149 and 150). In some cases such as 129 (batteries), the time-series equation incorporates technological assumptions for which there is no evidence in the survey data.

The result of these exclusions is a list of 76 sectors in which consumption is to be projected by new equations based entirely on the survey data. Of these, 25 are subsequently adjusted and 51 are not. For 7 sectors, both the diary and the interview information are needed in order to include all components, so the two equations are added together. Six other sectors were estimated from both survey sources; among these, the interview data gave better results five times and the diary equations were not used; the reverse was true for the remaining sector (diary data only).

Interpreting the Demand Equations: Income Effects

The linear coefficients β_1, β_{1h}, and β_{1a} (of permanent income, food at home and food away from home, respectively) are nearly always positive, with only two exceptions for β_1 and three for β_{1a}. Since the regressions should predict expenditure near zero when income falls toward zero, the constant (β_0) and log-inverse (β_2) coefficients may be expected to be of opposite sign. This is nearly always the case in the interview sectors, where $\beta_0\beta_2 > 0$ occurs only four times in 318. In the diary equations there are two terms β_{2h} and β_{2a} to compare to β_0, and they may be of different signs: if the product $\beta_0(\beta_{2h} + \beta_{2a})$ is tested, it is positive only 33 times in 162, and almost half of those occurrences are for single-person households, individuals who tend to eat away from home more often than larger households and to spend relatively little on foods for home consumption, the most important diary sectors.

If the regression specification is correct, and expenditures on particular sectors reach saturation as shares of income, there is no reason for the linear coefficients to differ by household size. The same marginal budget shares will eventually be reached by all households, and differences in their speed of approach to those constant shares will be reflected in different coefficients β_0 and β_2. Significant differences in β_1 across size classes can therefore be interpreted to mean either that consumption patterns do not converge to constant shares, or that the sample does not reach to high enough incomes, for one or more size classes, to allow the asymptotic marginal shares to be estimated accurately. This is particularly likely to be the case for large households, even with some allowance for economies of scale: if a two-person household reaches a constant marginal budget share in a given sector at an income Y^*, a six-person household will probably need an income of at least $2Y^*$ and perhaps as much as $3Y^*$ before converging to the same share. For sectors in which the marginal share is rising, therefore (luxuries), β_1 may be larger for large than for small households at comparable levels of total income (not per capita income), while the reverse should be true for sectors whose share is falling (necessities), even if the eventual saturation share is the same for all households. Luxuries in turn should be identified by $\beta_0 > 0 > \beta_2$ (reaching saturation from below) and necessities by $\beta_2 > 0 > \beta_0$ (saturating from above). Systematic identification will not be possible, however, if in the sample a category appears to be a luxury for one size class but a necessity for another class, because of differences in per capita income. The difficulty of interpretation is compounded by the fact that most of the sectors take very small shares of the average and marginal budget, with coefficients β_1 close to zero: β_0 and β_2 may reverse sign between size classes almost randomly, with substantial errors on both the income coefficients. Most empirical work with consumption functions that exhibit saturation in the budget share, in contrast, divides expenditure into no more than five or ten categories.

The interview equations show food expenditure to saturate at very low marginal shares (in excess of one percent, for food at home, only for families of three to five members). Except for four-person households, the marginal share is always higher for food away from home than for food at home; total food takes typically three to six percent of the marginal budget. Among non-food sectors, β_1 exceeds one percent only for 45 (furniture), 133 (automobiles), 147 (jewelry and silverware), 155 (air travel), 158 (telephone—for individuals only), 165 (credit agencies), 166 (insurance), 168 (real estate), 169 (lodging places), 170 (personal and repair services), and 174 (amusements).

Among the diary sectors, there is virtually always a significant income effect with respect to food at home (β_{1h} distinct from zero), but the coefficient for food away from home (β_{1a}) is often not significant. By far the largest effect is for sector 23, meat products, with coefficients of about 0.4 to 0.6. There are also substantial effects (β_{1h} of 0.05 or more) for sectors 7 (fruit and vegetables), 24 (dairy products), 25 (canned and frozen goods), 27 (bakery products) and 67 (cleaning and toilet preparations). It appears that the distinction between food at home and food away from home is useful for explaining expenditure in the detailed food sectors and is preferable to the use of total food expenditure as an "income" variable.

We also tested the expectation that when β_2 is negative, β_1 should increase with household size, and should be inversely related to size when β_2 is positive. Sectors 35 (fabrics), 39 (clothing), 73 (rubber products), 77 (leather products), 130 (engine electrical equipment), 152 (buses), 155 (airlines), 158 (telephone), 162 (water and sewer) and 168 (real estate) show this pattern among the interview equations, but there are also some counterexamples and many sectors for which no relation can be discerned. No pattern is evident in the equations from the diary data. It does not seem possible to conclude anything about whether the true long-run saturation share is independent of household size.

Equality among size classes of the income coefficients was tested by whether the coefficients were statistically distinguishable. This is, it should be emphasized, a very conservative measure of the extent of differences among households classed by size: two coefficients may have such large standard errors that they cannot be distinguished, yet differ enough in numerical value to have large consequences when used for projection. Among the interview sectors there is no household-size difference in the income coefficients for sectors 37 (miscellaneous textiles), 81 (cement), 116 (machinery), 122 (welding apparatus), 131 (x-ray equipment) and 142 (measuring devices), almost no difference for sectors 49 (paper), 73 (rubber products), 77 (leather products) and 157 (freight forwarding) and only occasional differences in several other sectors. At the other extreme, the income coefficients are demographically sensitive for sectors 38 (knitting), 39 (apparel), 40 (household textiles), 53 (periodicals), 124 (lighting and wiring), 133 (automobiles), 147 (jewelry), 152 (local road transport), 155 (airlines), 169 (lodging places), 170 (personal services), and 174 (amusements). These are sectors which might be expected also to show effects of household composition.

Among the diary sectors, there is a striking tendency for all households of two or more people to have similar income effects (indicated by β_{1h}), but for one-person households to be different. This is true of both

food (2, 26, 31) and nonfood (56, 57, 82, 87) sectors. Since households
of this kind are projected to increase more rapidly than any other kind,
demand projections could be substantially affected. There are few or no
demographic differences in only a few sectors: 34 (tobacco), 37 (textiles), 40 (household textiles) and 124 (lighting and wiring), most of
which were also estimated from the interview data. Strong demographic
effects are apparent for major food sectors (7, 23, 25 and 28), which is
not surprising. It may be noted, finally, that the log-inverse coefficients are usually indistinguishable among size classes.

The luxury-necessity distinction should appear in the sign of the
log-inverse parameter β_2, being negative and positive for these categories, respectively. Of the 54 interview sectors, 16 appear consistently
to be luxuries across all size classes. Another ten sectors appear to be
luxuries for households of six or more, and to have become necessities
for all smaller households. In another five cases, the sector is a necessity only for one-person households and a luxury for all larger units.
There are also four intermediate cases where households up to two or three
members regard the sector as a necessity and it is a luxury at larger sizes. These findings are consistent with the hypothesis that every sector
tends to change from a luxury to a necessity as per capita income rises,
and with the observation that in the sample, size and per capita income
are inversely correlated. At the income levels likely to prevail over the
projection period, therefore, the distribution of households by size can
be expected to determine whether many categories appear to be predominantly
luxuries or necessities, or--for those sectors which are still luxuries
to all size classes--how rapidly they come to seem necessary. Even if the
long run or saturation budget shares are identical for all households, the
household size distribution is important because of the curvilinear term
in the equations.

This distinction is more difficult for the diary sectors because of
the more complicated relation of expenditure to income. Most sectors appear as luxuries relative to total spending on food for home consumption,
using the sign of the coefficient β_{2h} as the test; but since food at home
is always a necessity relative to income, these sectors may in fact be
necessities. Of 29 sectors for which equations were estimated, eight show

$\beta_{2h} < 0$ for all six size classes, and another nine have a negative coefficient for all but one-person households, this being the group that eats
away from home the most, so that the relation to food at home is least
informative.

Age Effects

The effects of age on expenditure, given household size and composition, are indicated by the coefficients λ_{is} of the age dummy variables,
comparison being made to the median age class (35-44 years). As with the
income coefficients, it may be of interest to know whether these parameters differ among household sizes, but there are so many more coefficients (seven age classes for each of six household sizes) that it is
hardly possible to undertake all the comparisons. The principal question
is whether, for a given sector, there is a significant amount of age variation showing the same pattern at all sizes; if there is a distinct age
pattern for each size class, it becomes impossible to say anything about
the effects of aging in the population (which is a phenomenon modeled in
the basic demographic projections) without making detailed and somewhat
arbitrary assumptions about the joint distribution of age and size.

Once household size is taken into account, the age of the head has
little apparent further effect on expenditure, except for food at home and
some of its components. There are no significant age differences, or differences between coefficients in only one household size class, for 15 of
the 54 interview sectors, and there are frequent age differences for only
three sectors: 73 (rubber products), 166 (insurance) and 168 (real estate). Among the diary sectors, age differences are frequent only for
sectors 7 and 34 (fruits and vegetables, and tobacco), and there are few
or no differences in five sectors. Only one of the latter is a food sector (24: dairy products), and in general it appears that age has more
effect on food than on nonfood consumption. There is some tendency for
coefficients to differ more often for more distant age groups, but it is
not very pronounced. There appears also to be a slight tendency for differences to be most marked in the middle of the age distribution: that

is, the age group 30-34 or 35-44 is more likely to have a coefficient different from those of adjacent groups than is the case with a younger or older group.

The analysis just described is a useful summary of demographic sensitivity in the regressions, since it indicates how often two age groups show a <u>statistically</u> significant difference in expenditure for a given income and household size and composition. They say nothing, however, about whether the differences are <u>economically</u> significant; that is, whether they amount to a substantial fraction of expenditure in the sector. As with the income coefficients β_1 and β_2, two age coefficients, λ_i and λ_j, may be indistinguishable because of large standard errors, yet differ enough in numerical value that a change in the age distribution would markedly affect spending. We can test whether differences among age coefficients are economically important by comparing them to mean expenditure in the sector. To isolate a "pure" age effect, we suppose that income, household size and composition do not change. The appropriate comparison then is the mean of differences $\lambda_i - \lambda_j$ taken over the six size classes, using relative frequencies in the classes as weights, compared to the mean of expenditure, also over all size classes. That is, an age difference is economically important for a sector if the statistic

$$100 \cdot \frac{\Sigma_s \, (\lambda_{is} - \lambda_{js}) N_s}{\Sigma_s \, \overline{C}_s \, N_s} = \Lambda_{ij}$$

is large, where N_s is the number of households of size and $\overline{C}_s$ is average expenditure in that size class. Of course, Λ_{ij} may be small because λ_i exceeds λ_j in one size class but the reverse is true in another; Λ_{ij} will be large only if there are no such inconsistencies in the <u>pattern</u> of coefficients.

The values of Λ_{ij} were calculated so that positive values mean higher expenditure at younger ages, or spending declining with age, while negative values mean that spending increases with age. Very large values occur when mean spending is close to zero, so that even a small difference between coefficients has a relatively large impact. Sectors can be characterized by the typical size and sign of Λ, as follows:

		Size of Λ_{ij}	
Expenditure	Small (< 10%)	10-100%	Large (> 100%)
Declining with age	72	23, 31, 45, 54, 74, 97, 116, 123, 125, 126, 130, 133, 139, 145, 146, 147, 148, 165, 171	25, 49, 142, 157, 168
Rising with age	24, 40, 76, 77, 124, 174	3, 27, 28, 32, 33, 35, 43, 52, 53, 56, 59, 61, 73, 81, 140, 151, 154, 155, 162, 166, 170	7, 26, 37, 57, 67, 73, 170, 177
Non-monotonic (or no pattern)	38, 39, 40, 74, 82, 87, 128, 152, 158	29, 34, 37, 49, 58, 80, 103, 122, 124, 131, 153, 169	30

There is little or no systematic age effect for dairy products, textiles, clothing, leather and plastic products, tires, stone, aluminum foil, electric and electronic goods (except appliances), telephone service and amusements; at the other extreme, age differences are sometimes much larger than mean spending for fruits and vegetables, canned and frozen foods, grains, alcohol, soft drinks, some textiles, paper, cleaning and toilet preparations, rubber products, real estate services, personal and repair services, and postal service. Among the sectors for which age effects are typically neither very large nor very small, rising, declining, and non-monotonic age trends are all observed. There is some tendency for purchases of durable and household goods to decline with age, and for expenditure on services (including transportation) to rise. All three patterns are observed among food sectors.

Household Composition Effects

These effects are analyzed in almost the same way as those due to age. The first step is to compare the coefficients corresponding to different numbers of children and count the number of times that two coefficients are indistinguishable. The analysis differs in two ways from that of age effects. The first is that the number of children in a household is, at most, one less than the total number of members. No compositional differences exist therefore for one-person households, who are always adults; only the variable for one child, with coefficient κ_1, is intro-

duced into the regressions for two-person households, and so on. As a
result, the maximum number of equalities in the matrix of pairwise comparisons is not uniformly six, as with age coefficients, but varies according to the maximum number of children allowed.

The second difference is more subtle, and involves the proportions of
adults and children in the household as a function of household size. Consider only the comparison between zero children and one child. For a two-
person household, replacing an adult with a child means a 50 percent reduction in the number of adults, so it is to be expected that expenditures on
a great many sectors would change significantly. For a five-person household, however, such a replacement means only a 20 percent reduction in the
number of adults, and the child represents only one-fifth of the household
instead of one-half. Expenditures should therefore change less. For this
reason, it would be preferable to have the coefficients correspond to equal
proportional rather than absolute changes in the composition of the household, but that specification would be much harder to relate to changing
numbers of adults and children in the population. As an empirical matter,
the difficulty with the specification used here may not be very important,
because large households tend to consist predominantly of children; units
with one child and a large number of adults are rare.

Before taking up the analysis of the final specification of household
composition, we consider a crude test of overall importance of composition,
which was performed for the food expenditure sectors, using a trial specification in which children were distinguished by age into three groups.
The equation for two-person households then includes three dummy variables
rather than one, as in the simple adult/child specification adopted later.
Regressions were estimated with and without the appropriate block of dummy
variables for each size class, and the significance of the entire block
was tested. Composition effects were significant for some sectors in each
size class, but not always for the same sectors, nor did the number of
sectors affected show any relation to household size. It appears that
once age, size and income are taken into account, composition has little
effect on food expenditures, but this is not too surprising: food is consumed by all members of a household, and adolescent members, who are still
counted as children if aged 17 or less, may easily consume as much as

adults. Compositional effects are more likely to show up in sectors consumed predominantly by adults or predominantly by children (alcoholic beverages are perhaps the only example among foods).

The identical test was performed for the interview sectors, except that only households of three members were examined (where family composition is specified by six dummy variables, two for each of the three age-of-child classes). Family composition was significant (by an F-statistic) for 18 of the 54 sectors. The test was repeated using observed income and observed total expenditure (and their respective log-inverses) in place of the income variable Y^*, and in both cases 18 sectors showed significant composition effects: 15 sectors appeared to be affected in all three tests. The variables total food, food at home and food away from all consistently showed significant effects of household composition on spending.

The sectors where composition appears to be important include knitting, apparel and footwear (38, 39 and 76), as well as toys and sporting goods (148), in all of which children can be expected to affect expenditure. Children may also have fairly direct effects on a households' purchases of paper products (49), insurance (166), hotels and lodging (169) and personal and repair services (170), and on the tendency to move residences and thus pay real estate brokers (168). In some other sectors, the absence of children appears to raise expenditure, but what is observed is really a life-cycle effect: households tend to buy appliances (123), phonographic and photographic equipment (126, 145) in the early years of marriage, before children are born. It does not necessarily follow that a higher birth rate would reduce spending in these sectors or that a lower rate would increase it. The remaining affected sectors (books, rubber products, welding apparatus, x-ray equipment, trucking) bear no obvious relation to family composition or life cycle.

To return to the tests of our final specification, the summary analysis shows that for a given household size, the different possible adult-child combinations are not usually distinguishable: equality of coefficients tends to occur as often as is possible. There are, of course, some exceptions to this pattern. One is that compositional differences are important for a few sectors, such as 72 (tires), 73 (rubber products) and 34 (tobacco), although such differences are not observed in food sectors or in those for household durables. A second effect is that the coefficients for numbers of children are often indistinguishable from one another, but separately distinguishable from zero. This is equivalent to saying that it matters, for the determination of spending, whether a household contains any children, but it does not matter, or matters very little, just how many children there are. Finally, as is to be expected, coefficients are more likely to differ, the greater is the difference in the number of children represented. This effect, however, is much weaker than the distinction between some children and no children. The importance of this separation may validate the decision to project population separately in households with and without children, while suggesting that the exact distribution of children among the former group of households is of much less importance.

As with the effects of age, it is necessary in analyzing household composition to distinguish between statistical and economic importance. The latter is evaluated by a statistic of the same form as the Λ_{ij} calculated for age differences; that is,

$$100 \cdot \frac{s > \Sigma_{\max(i, j)} (\kappa_{is} - \kappa_{js}) N_s}{\Sigma_s \overline{C}_s N_s} = K_{ij}$$

where κ_{is} is the coefficient for i children in size class s, and the restriction on the sum in the numerator means that only households large enough to have i or j children are included (there is no restriction on the weighted mean expenditure in the denominator, so that all coefficient differences, irrespective of the household size classes involved, can be compared to the same level of spending). A positive value of K means that expenditure is higher when there fewer children (or more adults), and a negative value means that expenditure rises with the number of children. Very large values of K reflect mean expenditures close to zero. Provided the effect of adding children to the household is monotonic, the absolute value of K should increase as i and j differ more; this tendency is observed sometimes but not invariably.

The values of K can be characterized, for each sector, by their typical size and sign as follows:

Size of K_{ij}

xpenditure	Small (< 10%)	10 - 100%	Large (> 100%)
ecreasing ith children	77	23, 54, 72, 80, 97, 124, 130, 133, 142, 151, 155, 166	7, 30, 32, 34, 50, 67, 73, 152, 157, 169, 177
ncreasing ith children	24, 38, 39, 40, 52, 82, 87, 158	2, 25, 37, 40, 43, 45, 49, 53, 56, 59, 73, 81, 103, 122, 125, 131, 140, 146, 148, 153, 168, 170, 174	26, 28, 37, 124
n-Monotonic or no pattern)	27, 35, 74, 76, 123, 162	29, 31, 33, 49, 61, 74, 116, 126, 128, 139, 145, 147, 154, 165, 171	57, 170

There are only very small compositional differences, relative to mean expenditure, in dairy and bakery products, clothing and textiles (including leather), stone, aluminum, appliances, telephone, and water and sewer services. At the other extreme, the composition of the household makes a relatively large difference for fruits and vegetables, flour, sugar, alcohol, fats and oils, tobacco, cleaning supplies, rubber products, lodging places, local transportation and a few other goods and services. With the exception of alcohol and tobacco, these are not sectors bought exclusively for consumption either by adults or by children.

To a large degree, the same sectors show large compositional effects that show large age affects: 15 sectors are so classified in each case, and 10 of these are the same in both cases. Small coefficient differences also appear in the same sectors for both variables: 31 sectors are so classified in one or the other case, and 10 are the same in both cases. This association is probably due to the considerable degree of correlation between age and household composition given the number of members. If the equations were not estimated separately by size class, the association would probably be much weaker. Expenditure declines as there are more children (or increases with the number of adults) not only for clearly adult purchases such as alcohol and tobacco, but for meat, fruits and vegetables, fats and oils, books, automobiles and tires, rail and air transport, hotels, postal services and a variety of machinery and equipment. Children lead to increased spending on poultry and eggs, canned and frozen foods, flour, sugar, furniture, appliances, toys, personal and repair services, amusements, and a variety of sectors associated with home-owning, such as fertilizer, cement, farm machinery, wiring, and real estate services.

Summary

Household size has been shown to have an important effect on spending, since income coefficients frequently differ among size classes. We have not tested whether the coefficients of age and composition also differ among size classes, but it is likely that they do, and this increases the importance of classifying households by size. The other two demographic variables--age of the head and relative numbers of adults and children--do not often show statistically distinguishable differences among classes, but the coefficient differences are large relative to mean expenditure, implying substantial relative differences among households, for about one-fifth of all sectors, some of which are appreciable shares of total expenditure. Smaller but still notable age and compositional differences--for example, differences of the order of 30 percent of mean spending--characterize many more sectors.

We do not try here to say which effect or which variable is most important, partly because some of the equations are subsequently modified, but mostly because so many coefficients are involved that it is virtually impossible to summarize the four effects--of size, age, composition and income--simply by looking at the equations derived from the survey data. It becomes much easier to measure and separate those effects after the expenditure equations have been combined with the household and income projections to estimate a single value of spending for each sector in each scenario. This evaluation was therefore presented for large categories in chapter 1 and is repeated at the level of individual sectors in chapter 6.

Chapter 3

EDUCATION: CURRENT AND SCHOOL
CONSTRUCTION EXPENDITURES

The object in studying this component of demand was to make long range forecasts of total national expenditures for education which would be sensitive to the age composition of the population and to the rate of economic growth, but not necessarily to the structure of individual households or the distribution of income among them. We wanted to include all age groups in the population, and to treat public and private education together so as to avoid the inconsistencies that could arise if demands for the two were projected separately. The bulk of education is publicly provided and therefore is not part of personal consumption expenditure.

The model anticipates no changes in public policies or programs that would significantly affect enrollment rates, per-pupil expenditures, or the relative importance of public and private schools. Apart from changes in tastes or in teaching productivity, certain exogenous political events could, of course, make this assumption unrealistic. The first event is a "tax revolt" which would limit financing of public schools, making it extremely difficult to expand expenditures on education as called for in our projections, although probably not affecting enrollment rates appreciably. The second possible event is the establishment of substantial tuition tax credits for private elementary, secondary and college tuition. This could bring a shift in enrollment back to private schools. It would be a reversal of current trends, although it would not have a great effect on total educational spending, private schooling--at most levels--being only a little more expensive than public.

The model projects school construction expenditures and non-salary current expenditures directly, in dollar terms. Expenditure on personnel, however, is projected by first projecting requirements for staff and then applying an assumption about salaries. Education is the only category of expenditure for which labor costs are projected separately from non-labor costs. This is because "productivity" is approximated by student-teacher ratios, which directly affect costs per pupil.

Background: Previous Models

Although the objective outlined above is quite straightforward, it appears not to have been attempted before. Other education models which we have studied are deficient for our purposes in one or more respects, although they may contain useful components. Several previous models are discussed below.

Census Bureau Enrollment Projections

The Bureau of the Census periodically makes long-range projections of school enrollment. The most recent projections were published in 1972 and extend to the year 2000 (Bureau of the Census, 1972). In general, the methodology is to present three alternative projections of enrollment rates, one of which is a continuation of historical rates of change (from 1950 to 1970), the second of which reflects a rate of change about half as great as the first, and the third of which assumes constant 1970 enrollment rates.

This approach is satisfactory for the levels of education that are close to saturation (100 percent), which are elementary and secondary. It would not be worthwhile to try to link the small changes that may be expected in those enrollment rates to economic growth. College enrollment rates are below fifty percent, however, and it would be desirable to link them to some measure of economic growth rather than simply imposing a time trend.

National Center for Education Statistics

The NCES publishes a ten-year projection of school enrollment, instructional staff, and educational expenditures every year (NCES, 1977). Enrollment rates are projected in considerable detail: by grade level for grades 1 through 12, and by various kinds of post-secondary institutions and types of enrollment. The enrollment rates are projected as a function of time by fitting historical data to a linear function, or to a logistic curve if enrollment is already near saturation. Expenditures are projected by multiplying the projected enrollments by projected per pupil expenditures which are obtained from time trends. This method also does not satisfy our requirement that post-secondary enrollment rates and expenditures be related to economic growth.

Population Commission

The education forecasts which were made for the Population Commission (Butz and Jordan, 1972) include considerable demographic detail. Enrollments are projected to the year 2000 for four different levels of schooling, from pre-kindergarten to higher education. For each level two enrollment rate projections (high and low) are presented, both of which are obtained by trend extrapolation or by arbitrary assumption.

The expenditure forecasts are presented for five separate categories of expenditures: professional instructors, paraprofessional instructors, physical plant, equipment and instructional resources, and auxiliary services. The forecasts are derived by estimating the expenditures in different categories that would be required for a somewhat more costly hypothetical future school at each level, as well as for a typical school in 1970. Then three alternative forecasts are made. The first assumes a continuation of the 1970 level of expenditures per pupil. The second assumes that by 2000, 30 percent of all schools would be spending at the higher hypothetical future level. The third assumes that 80 percent of schools would be of the more expensive future type in 2000. The scope and level of detail in this model are excellent, and the forecasts may be quite reasonable. However, the way in which they were derived does not fill our need for a forecast that is linked to economic growth, since the transition to higher-cost and presumably higher-quality schools is independent of income changes.

INFORUM

Almon's INFORUM model (Almon, Buckler, and Reimbold, 1974) forecasts educational expenditures by means of three regression equations. One equation, for private schools and other nonprofit organizations, projects per capita expenditures as a linear function of per capita disposable income and time. The second equation projects non-labor public school expenditures per person aged 6 to 19 as a linear function of per capita disposable income. The third equation projects the number of public school employees per person aged 6 to 19 as a function of per capita disposable income.

Almon's equations have the advantage of being directly linked to income growth. They have three disadvantages, however, which make them unsuitable for our purposes. There is not enough demographic sensitivity in taking the total population aged 6 to 19 as the school-age population, and certain groups, such as preschool and older college students, are excluded. There is also no link between public school expenditures and private school expenditures, so that one sector may not properly complement the other; total enrollments are not controlled. Finally, private educational spending is combined with some types of non-schooling expenditure on non-profit institutions.

Choice of Approach

Having found deficiencies in existing models, we had to choose a hybrid approach for an original model. Cross-section data are clearly not appropriate in all respects; we are interested in the relationship between educational expenditures and income, and in the cross-section there would not be much relationship to individual household incomes due to the importance of public education, particularly for elementary and secondary schools. Time series data, however, should yield sensible results as long as past behavior has not been fluctuating or erratic and as long as we are careful not to exceed reasonable upper and lower limits. To use time-series data means forgoing any connection between school attendence and household composition, however; we must assume that the likelihood (and the cost) of school attendance do not depend on a person's family size or composition, and therefore total enrollments and expenditure do not depend on the distribution of population or of income among households. Again, this is reasonable for primary and secondary school but may be wrong where college education is concerned.

Availability of Data

Fortunately, there is an abundance of time series data available for virtually all determinants and components of educational expenditures for the past thirty or forty years. School enrollment by age is published annually in Series P-20 of the Current Population Reports. Fulltime-equivalent enrollment for higher education for the previous ten years is published every year by the National Center for Education Statistics in their Projections of Education Statistics series. Expenditures per pupil can be calculated for elementary and secondary schools from data in Statistics of State School Systems by the NCES. For higher education it can be found in the annual NCES Digest of Education Statistics and the older Biennial Survey of Education in the United States. Numbers of educational personnel can be found in the same sources as expenditures.

General Structure of the Model

We show, in figure 3-1, a flowchart of the calculations and data sources for current spending at one level of education. The sequence described below is repeated for each level of education.

(1) The enrollment rate is projected as a function of time or per-capita GNP and (for college) other variables.

(2) Enrollment is calculated as the product of the enrollment rate and the population in the appropriate age group.

(3) Fulltime-equivalent enrollment equals enrollment times the ratio of fulltime-equivalent to total enrollment. This ratio is constant for all levels below college, and varies with time (but not with income) for college enrollment.

(4) Per-pupil current expenditures for non-labor inputs are projected as a function of GNP per capita only.

Figure 3-1. Summary Diagram of Projection Model for
Current Expenditure, One Level of Education

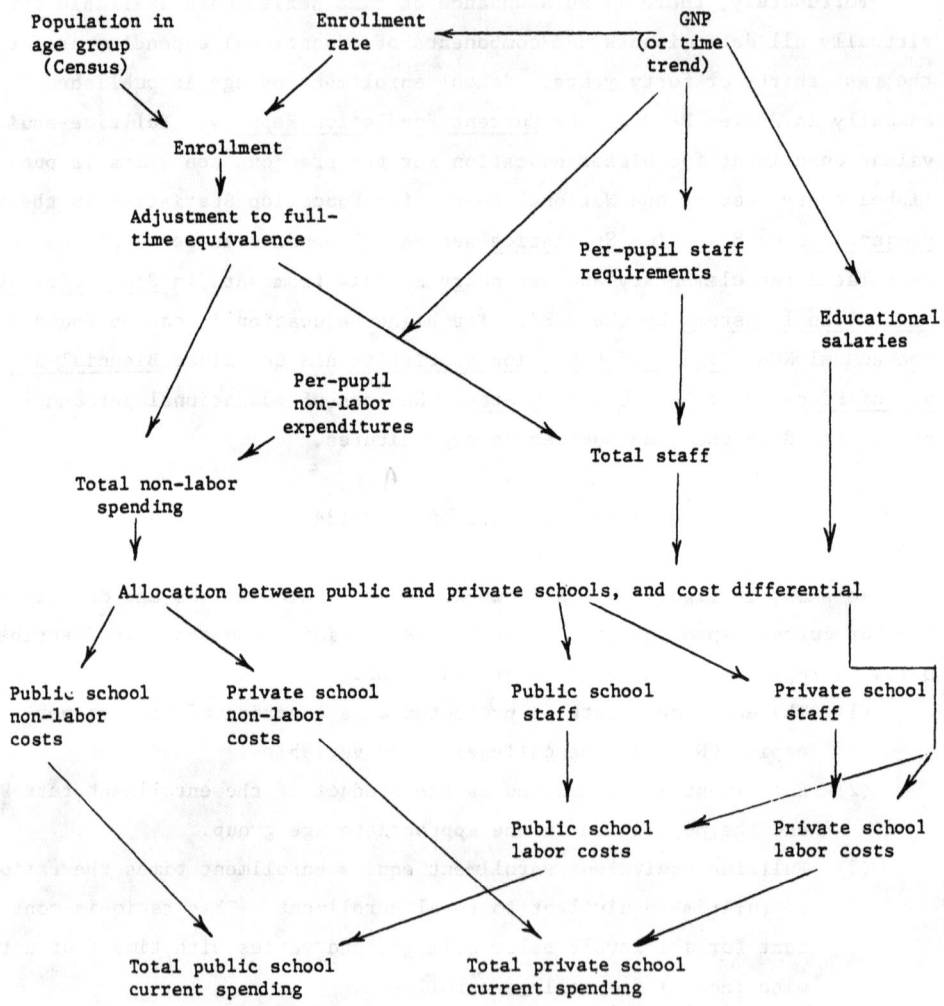

```
Population in          Enrollment                    GNP
age group                rate                      (or time
(Census)                                            trend)

              Enrollment

        Adjustment to full-
        time equivalence
                                        Per-pupil staff
                                         requirements
                                                              Educational
                                                               salaries
            Per-pupil
            non-labor
            expenditures
                                          Total staff
   Total non-labor
   spending

Allocation between public and private schools, and cost differential

Public school      Private school     Public school     Private school
non-labor          non-labor          staff             staff
costs              costs

                                     Public school      Private school
                                     labor costs        labor costs

   Total public school        Total private school
   current spending           current spending
```

98

(5) Per pupil non-labor expenditures times full time equivalent enrollment of public and of private schools equals public and private school current non-salary expenditures, respectively.

(6) School personnel requirements are projected by projecting faculty-student ratios, as functions of GNP per capita for pre-college levels and as a function of time alone for college. These ratios--doubled to account for non-teaching staff--are then applied to fulltime-equivalent enrollments.

(7) Salary expenditures are the product of school personnel and a wage rate, which is constant if labor productivity does not grow in the economy, or if relative wages in education are assumed to decline.

(8) Salaries plus non-salary expenditures equal current expenditures

Age Groups and School Levels

We have made the simplifying assumption that, with one exception, each age group is enrolled in only one level of school. Table 3-1 below shows the age groups used and the levels of education to which they correspond in the model.

Table 3-1. Age Groups and Levels of Education

Age Group	Level of Education
3-4	Prekindergarten
5-13	Elementary (including kindergarten)
14-17	Secondary
18-24	College undergraduate
25-34	College graduate
18-34	Special (trade, vocational, technical)
35+	College graduate (ignoring possible college undergraduate enrollment by older people)

We have not treated men and women students separately, for two reasons. First, the male-female ratio in the school-age population does not change appreciably through time. Secondly, there are no cost differences between male and female students. It does not matter, for our purposes, that

enrollment rates may be rising largely because of higher female enrollments in post-secondary education; the saturation rate is still 100 percent of the total age group.

Enrollments and Rates by Level

Elementary

The elementary school enrollment rate has fluctuated slightly in recent years, with a peak in 1971 of 98.71 percent. We have adopted the
Census Bureau's intermediate enrollment rate projection, which is 97.13
percent for 1980, and 97.3 percent for 1985 through 2000. Beyond 2000,
we have linearly interpolated between 97.3 percent in that year and 98.71
percent in 2025, assuming that a return to this peak value represents
saturation.

Pre-kindergarten

Due to the absence of historical data for pre-kindergarten enrollment
rates, they were projected by analogy with the past behavior of kindergarten enrollment rates, on the assumption that we are experiencing a
continuous process of the extension of schooling to younger and younger
children. The following linear regression was performed on kindergarten
enrollment rates from 1940 to 1976 as a function of time:

$$\text{Kindergarten enrollment rate} = 46.670 + 1.445\,(t - 1939), \quad R^2 = 0.873$$

where t is the year.

The calculated values from the regression were normalized to a set of
indexes, such that 1974 was equal to 1.0. These indexes were used, beginning with the 1974 pre-kindergarten enrollment rate, to obtain a set of
estimates running back in time which were fitted to a straight line to
yield the following equation:

$$\text{Pre-kindergarten enrollment rate} = 13.822 + 0.4279\,(t - 1939)$$

where t is the year.
A logistic curve might have captured better the acceleration in the early
1950s, but the linear fit is considered adequate. Table 3-2 shows the
data used and the results.

Table 3-2. Observed and Calculated Kindergarten Enrollment Rates
(Percentages) and Calculated Pre-Kindergarten Enrollment
Rates

Year	Observed kindergarten rates	Calculated kindergarten rates	Index, 1974=1.0	Calculated pre-kindergarten rates
1940	43.0	48.115	0.4948	14.250
1945	60.4	55.340	0.5691	16.390
1946	62.0			
1947	58.0			
1948	56.0			
1949	59.3			
1950	58.2	62.565	0.6434	18.530
1951	54.5			
1952	54.7			
1953	55.7			
1954	77.3			
1955	78.1	69.790	0.7177	20.670
1956	77.6			
1957	78.6			
1958	80.4			
1959	80.0			
1960	80.7	77.015	0.7920	22.810
1961	81.7			
1962	82.2			
1963	82.7			
1964	83.3			
1965	84.4	84.240	0.8663	24.949
1966	85.1			
1967	87.4			
1968	87.6			
1969	88.4			
1970	89.5	91.465	0.9406	27.089
1971	91.6			
1972	91.9			
1973	92.5			
1974	94.2	97.245	1.0000	28.800
1975	94.7			
1976	95.5			

Secondary

The enrollment rate projection for secondary schools was adapted from the Census Bureau's intermediate projection to the year 2000. After 2000 we assumed the same rate of growth in enrollment rates as during the period 1980-2000. Table 3-3 shows the projected enrollment rates for pre-kindergarten, elementary and secondary education, at five-year intervals.

College, Undergraduate

The undergraduate enrollment rate was hypothesized to be a function of the logarithm of per capita GNP, the log of the real price of college (using the weighted average tuition at public and private colleges to measure price, as discussed below), the percentage of the age group serving in the armed forces, the draft status of full-time college students, discharges from the armed forces as a percentage of the age group (the number of persons in the service three years earlier was used as a proxy for discharges), and time. A regression was run on data for 1957 through 1976. The time, price and armed forces-discharge terms were dropped due to insignificance or wrong sign,[1] and the equation was re-estimated in the following simpler form:

$$ERU = B_0 + B_1 \log Y_t + (B_2 + B_3 D_t) A_t$$

where ERU = undergraduate enrollment rate as a share of 18-24 age group

Y = per capita GNP in 1971 dollars

D = dummy variable for draft status
   1 = college deferment in war year
   0 = no deferment or no war

A = share of age group in armed forces.

The data used for this regression are shown in table 3-4. The coefficients (all of which are statistically distinct from zero) are shown below:

$$ERU = -1.596 + 0.494 \log_{10} Y_t + (0.233 D_t - 0.129) A_t, \quad R^2 = 0.955$$

---

[1] The price term had a positive sign, probably because of collinearity with time and income. Also, tuition is a rather poor measure of actual price since it does not take account of scholarships, student loans, and other forms of financial aid.

Table 3-3. Projected Enrollment Rates for Pre-Kindergarten, Elementary
and Secondary Education, 1975-2025 (Percentages)

Year	Pre-kindergarten	Elementary	Secondary
1975	29.2	97.77	92.71
1980	31.4	97.13	93.08
1985	33.5	97.31	95.17
1990	35.6	97.30	95.27
2000	39.9	97.30	95.11
2005	42.1	97.58	95.62
2010	44.2	97.86	96.14
2015	46.3	98.14	96.66
2020	48.5	98.43	97.18
2025	50.6	98.71	97.71

College, Graduate

The enrollment rate for graduate students was projected in a manner
similar to undergraduates. The initial regression specification was the
same; in this case, however, the price term was found to be significant
and was retained. The data for this equation are shown in table 3-5. The
final regression equation is shown below (all coefficients significant):

$$ERG = -0.57 + 0.228 \log_{10}Y_t - (1.084 + 0.35\ D_t)A_t - 0.066 \log_{10}P_t,$$
$$R^2 = 0.967$$

In order to use this equation for forecasting, we required a forecast of
the price variable P which represents average tuition. We have done this
by assuming that tuition will follow the same pattern as college costs.
We have divided college costs into a salary portion (82 percent of the
total) and a non-salary portion (18 percent of the total). The non-salary
costs are assumed to remain constant, consistent with the constant-dollar
assumption in the rest of the model. The salary costs, on the other hand,
are assumed to increase at the same rate as average labor productivity
because the education sector, even though its productivity may remain
constant, must compete for labor with the rest of the economy where productivity is increasing. This forces labor costs to rise in the education
sector, as in other service sectors where productivity is constant. Par-

Table 3-4. Data for Regression Model of College Undergraduate Enrollment Rates, 1957-1976

Year	Enrollment Rate (Percent)	Per Capita GNP (1971 Dollars)	Percentage of 18-24 year-olds in armed forces[1]	Draft status
1957	14.74	3714	10.9	0
1958	15.25	3645	9.5	0
1959	15.33	3800	8.4	0
1960	16.11	3826	7.8	0
1961	16.87	3857	7.3	0
1962	18.31	4018	8.4	0
1963	18.79	4117	7.8	0
1964	19.02	4274	7.7	1
1965	22.08	4470	7.1	1
1966	22.92	4682	8.6	1
1967	22.84	4757	9.6	1
1968	23.41	4916	9.7	1
1969	24.62	4993	9.3	1
1970	23.52	4923	7.6	1
1971	24.09	5018	6.2	1
1972	24.16	5260	4.8	1
1973	22.95	5499	4.5	0
1974	23.47	5391	4.1	0
1975	25.12	5281	3.8	0
1976	25.50	5559	3.6	0

Draft Status: 1 = College deferment in war year; 0 = no deferment or no war

[1] Male servicemen 18-24 as percentage of total population 18-24.

Table 3-5. Data for Regression Model of College Graduate Enrollment
Rates, 1957-1976

Year	Enrollment rate (Percent)	Per Capita GNP (1971 dollars)	Price (1971 dollars)[1]	Percentage of 25-34 year-olds in armed forces[2]	Draft status
1957	3.09	3714	498	2.8	0
1958	3.14	3645	531	2.8	0
1959	3.14	3800	563	2.9	0
1960	3.27	3826	582	2.9	0
1961	2.89	3857	618	3.1	0
1962	3.28	4018	601	3.5	0
1963	3.23	4117	633	3.4	0
1964	3.48	4274	606	3.2	1
1965	4.14	4470	680	3.1	1
1966	4.11	4682	695	3.1	1
1967	4.59	4757	675	3.1	1
1968	4.85	4916	660	3.2	1
1969	5.48	4993	681	2.9	1
1970	5.33	4923	660	2.5	0
1971	6.17	5018	676	2.4	0
1972	6.42	5260	706	2.2	0
1973	6.39	5499	689	2.1	0
1974	7.40	5391	638	1.9	0
1975	7.99	5281	635	1.9	0
1976	7.77	5559	670	1.8	0

Draft status: 1 = Graduate student deferment in war year
               0 = No deferment or no war

[1]Data on graduate tuition were not readily available, so undergraduate
tuition was used on the assumption that the rate and direction of change would
be roughly the same.

[2]Male servicemen aged 25-34 as percentage of total population aged 25-34.

tially offsetting this upward cost push is the trend toward declining
faculty/student ratios in higher education, which is shown historically
in table 3-10, below, and has been assumed to continue to decline. The
faculty/student ratio was indexed so that 1975 equals 1.0. This index
was used as a multiplicative adjustment factor to the salary portion of
college costs. The complete forecasting equation for college tuition
then is:

$$P_t = 0.18\ P_{1975} + 0.82\ P_{1975}\ (FSI_t)\ (PROD_{t-1975})$$

where P = college tuition
PROD = annual productivity increase
FSI = index of the faculty/student ratio
t = year

Special Schools

The enrollment rate for trade, technical and business schools has
hovered around two percent of the population aged 18 to 34 for the last
few years, following a decline from about three percent. These data are
shown in table 3-6.

Older Students

Although it is possible that the percentage of persons aged 35 or
more enrolled in school will rise in the future, there are not enough
historical data to determine the trend. We have used a constant term of
0.7 percent based upon the 1975 enrollment of persons aged 35 or more
(Current Population Reports, P-20, No. 303, 1976), adjusted for full-time
equivalence.

Calculation of Total Enrollment

Total enrollments are calculated by multiplying the enrollment rates
by the projected populations in the appropriate age groups, taken from the
Bureau of the Census Series D, E and F. Whereas enrollment rates may
differ according to income, population size and structure do not in this
model; thus, differences between the growth and no-growth scenarios depend
only on the enrollment rates.

Table 3-6. Enrollment in Special Schools as a Percentage of the
Population Aged 18 to 34

Year	Enrollment Rate
1962	2.85
1963	2.87
1964	3.09
1965	2.88
1966	2.50
1967	2.60
1968	2.65
1969	2.61
1970	2.53
1971	2.18
1972	2.62
1973	1.71
1974	1.93
1975	1.91

Fulltime Equivalence Adjustment

In order to obtain a firm basis on which to project educational expenditures, the projected enrollments were adjusted to a fulltime equivalent basis. The method of adjustment varied with the age group and educational level, as described below.

Pre-kindergarten
Since most 3 and 4-year-olds attend prekindergarten schools for a
half day, the full-time equivalent enrollment was assumed to be a constant one-half of total enrollment.

Elementary and Secondary
Since virtually all elementary and secondary school students attend
school for a full day, fulltime equivalent enrollment was assumed to be
equal to total enrollment for both of these levels.

Undergraduate and Graduate College Students

The percentage of college undergraduates who are fulltime students has been declining for a number of years. We have assumed that this decline will continue, at a decreasing rate, until 1990, after which the fulltime percentage of undergraduates will remain constant. We have used the same approach to project the fulltime percentage of graduate students. Some of the historical data and the projections are shown in table 3-7. A part-time student is assumed to be equivalent, on average, to 0.38 fulltime students (NCES, 1977).

The equation for calculating the fulltime equivalent enrollment of college undergraduates and graduate students is then

$$FTEE_t = E_t \cdot PF_t + 0.38E_t \ (1 - PFT_t)$$

where FTEE = fulltime equivalent enrollment

E = total enrollment

PFT = share of enrollment that is fulltime

Special Schools

Enrollment in special schools of persons aged 18 to 34 is assumed, in the absence of better estimates, to have the same relationship to fulltime equivalent enrollment as college undergraduates.

This completes the calculation of the number of pupils at each level in each scenario. Expenditures are projected on a per-pupil basis, and the multiplication by the number of students yields total expenditure by type and level.

Non-Salary Current Expenditures Per Pupil (NSCE)

Per pupil expenditures on all items except personnel, by level of education, have been projected by regressing historical data on constant dollar expenditures against per capita GNP in constant dollars (Y).

Pre-kindergarten Through Secondary Schools

Due to a lack of historical data on pre-kindergarten and kindergarten expenditures, they were assumed to have the same growth rate in the future

Table 3-7. Percentages of Undergraduate and Graduate College Students
Attending Full Time, 1965-1975 (Observed) and Projected,
1976-1990

Year	Undergraduates	Graduates
1965	73.5	36.7
1966	74.3	37.1
1967	74.1	37.3
1968	73.9	38.1
1969	73.2	38.1
1970	72.4	36.8
1971	72.1	38.3
1972	69.5	36.9
1973	68.8	36.4
1974	66.3	36.0
1975	65.0	35.9
1976	64.8	35.8
1977	63.0	35.7
1978	62.2	35.6
1979	61.3	35.5
1980	60.5	35.4
1981	59.7	35.3
1982	59.0	35.3
1983	58.4	35.2
1984	57.8	35.1
1985	57.1	35.0
1986	56.6	35.0
1987	56.0	34.9
1988	55.5	34.9
1989	55.1	34.8
1990 to 2025	54.8	34.8

Source: For historical data (through 1975), NCES, 1977.

as elementary expenditures. Statistical time series on elementary and secondary expenditures were obtained by taking data on combined elementary and secondary per pupil expenditures from the biennial Statistics of State School Systems published by the National Center for Education Statistics. These expenditures were allocated to elementary and to secondary pupils using information on per pupil expenditures by level of education in 1970 in the report of the Population Commission (Butz and Jordan, 1972, pp. 202-204). Table 3-8 shows the time series derived in this manner.

After trying several functional relationships between per pupil expenditures and per capita GNP, a logarithmic reciprocal function was selected because it gave a good fit and also yielded reasonable projections. This form leads to saturation in expenditure as income rises. The equation for elementary school is

$$\log_{10}\text{NSCE (Primary)}_t = 3.166 - \frac{3809.8}{Y_t}, \text{ and}$$

The equation for secondary school is

$$\log_{10}\text{NSCE (Secondary)}_t = 3.330 - \frac{3809.8}{Y_t}$$

The way of splitting costs between elementary and secondary schooling guarantees that the coefficients of $1/Y_t$ are equal in the two equations. The saturation levels are respectively \$1466 and \$2138, but by the end of the projection period only about one-half these levels are reached.

College Per-Pupil Expenditures

As with lower levels of schooling, data from 1940 to 1970 on non-salary current per-pupil expenditures were regressed against per capita GNP. The data are from the Biennial Survey of Education and are shown in table 3-9. The results of the regression are shown below. The dependent variable in this equation,

$$\log_{10}\text{NSCE (College)} = 4.081 - \frac{7325.1}{Y_t}$$

represents average costs for higher education. (The saturation level is \$12,050 but in 2025 the level reached is only about \$2,000.) In order to

Table 3-8. Per Pupil Current Non-Salary Expenditures on Elementary and
            Secondary Pupils, and Per Capita GNP, in 1971 Dollars,
            1940-1974

Year	Per Capita GNP	Per Pupil Expenditure	
		Elementary	Secondary
1940	2428	61.57	89.84
1942	3147	69.24	101.00
1944	3824	76.91	112.17
1946	3149	84.59	123.33
1948	3108	103.36	150.67
1950	3287	103.80	151.32
1952	3564	125.73	183.41
1954	3531	124.65	181.96
1956	3715	139.49	203.63
1958	3645	139.38	203.35
1960	3826	153.02	223.10
1962	4018	166.56	242.90
1964	4274	180.31	263.06
1966	4682	204.54	298.29
1968	4916	259.29	378.08
1970	4923	290.80	424.15
1972	5260	340.53	496.62
1974	5391	378.42	551.88

match the detail in the enrollment projections, these costs were allocated
to undergraduate and graduate level education according to the assumption
that graduate education is 2.5 times as expensive as undergraduate education (Carnegie Commission, 1972).

Per pupil expenditures on students aged 35 and over were assumed to
be the same as on graduate students aged 25 to 34 (which is an overesti-
mate for older people attending undergraduate school). Expenditures on
special school students aged 18 to 34 were assumed to be the same as on
college undergraduates.

Table 3-9. Non-Salary Current Per Pupil Expenditures on Higher
Education, 1940-1970, in 1971 Dollars

1940	$ 18
1942	24
1944	42
1946	34
1948	64
1950	96
1952	115
1954	123
1956	136
1958	169
1960	199
1962	233
1964	266
1966	322
1968	406
1970	526

Educational Personnel Requirements

In every other expenditure category or sector, it is assumed that
wages increase at the same rate as overall labor productivity. The change
in employment in a sector would then depend on the relative rates of
change of demand for the output and of productivity, while prices would
stay constant. In other sectors we do not attempt to model employment,
but in education we effectively measure inputs rather than output, and
thus project employment directly. Personnel requirements are assumed to
be insensitive to educational salaries; that is, there is no pressure of
higher prices leading to higher student/teacher ratios or other ways to
lower costs per student. Also, except for college enrollments, the
demand for education is insensitive to price or unit cost.

Future requirements for educational personnel were estimated by projecting the faculty/student ratio for each level of education and assum-

ing that total personnel are equal to twice the number of teaching faculty. This ratio was found, from historical data, to be increasing for elementary and secondary education but decreasing for higher education. The
historical data are shown in table 3-10. For this reason, regressions
against income were used for the lower educational levels, but judgmental
extrapolation was relied upon for post-secondary education in order to
avoid reaching absurdly low levels of faculty per student in higher education as income rises.

The following regression equations (with saturation) were found to be
best for elementary and secondary education.

$$\log_{10} FS(Ele)_t = 1.821 - \frac{901.254}{Y_t}$$

$$\log_{10} FS(Sec)_t = 1.903 - \frac{900.866}{Y_t}$$

A non-linear curve was selected for the college faculty-student ratio in
order to achieve a good fit to the data and a reasonable projection. The
resulting equation is shown below; it does not depend on per capita GNP:

$$FS(Coll)_t = 0.3276 \exp[-.0712(t - 1921.3)] + .0518, \quad t = year$$

The result of applying this equation is that the projected higher education faculty/student ratio declines from .0607 in 1975 to .0520 in 2025.
Faculty resources are allocated to undergraduate and graduate education in
the same fashion as non-salary expenditures, by assuming that graduate
education requires 2.5 times as many resources as undergraduate education.

Employee Compensation

Expenditures on employee compensation were calculated by multiplying
the projected requirements for educational personnel by the average 1971
wage for education employees, for the constant-income projections. For
the projections assuming income growth, the base (1971) wage was adjusted
upward at the same rate as labor productivity so as to keep educational
salaries competitive. Because salaries are such a large share of total
current costs, this is the major difference between the growth and no-

Table 3-10. Faculty/Student Ratios, by Educational Level, 1940-1974

Year	Elementary	Secondary	College
1940	.0339	.0409	N.A.
1942	.0338	.0408	N.A.
1944	.0338	.0408	.1315
1946	.0337	.0406	.0772
1948	.0339	.0409	.1041
1950	.0361	.0436	.1175
1952	.0341	.0412	.1185
1954	.0340	.0411	.0728
1956	.0348	.0420	.0665
1958	.0375	.0453	.0699
1960	.0425	.0513	.0674
1962	.0391	.0472	.0643
1964	.0395	.0477	.0621
1966	.0415	.0501	.0613
1968	.0445	.0537	.0692
1970	.0466	.0562	.0671
1972	.0475	.0573	.0627
1974	.0504	.0608	.0614

growth projections. (The other two differences are growth in non-salary costs as income rises, and increases in college enrollment rates when income is higher.) Both sets of projections effectively assume that the "output" of education is the passage of a student through a year of schooling, and that "productivity" is simply a matter of the faculty/student ratio. No allowance is made for improved procuctivity in the form of better teaching, more material learned by the student, etc. Because these assumptions have such a large effect on total spending, we also made a series of projections in which income and productivity in the economy were assumed to grow, but teachers' salaries remained at 1971 levels. These projections can also be interpreted as showing the course of salary expenditures if the cost per student were held (nearly) constant by reducing staff and using other means to maintain the productivity of education. The actual course of faculty compensation seems likely to lie between these two assumptions: some increase in real wages but also some slippage with respect to mean income in the economy.

Public and Private Education: Enrollment and Personnel

For each level of education, projections have been made of the public and private shares of enrollment. These projections are based on continuations of recent trends through 1985 and constant shares thereafter. Table 3-11 shows the historical and projected shares for elementary, secondary and higher education.

Educational expenditures and personnel requirements were allocated to public and private education by assuming that private education requires 35 percent more resources per pupil than public education at the post-secondary level, and essentially the same amount of resources per pupil at other levels of education (elementary, secondary and special education). Private post-secondary schooling is assumed to use more of all resources, without changing the composition of inputs. This is partly a matter of higher faculty/student ratios, and may also depend on the retention rate of students. No changes are forecast in the type of student body at either type of school.

School Construction

As with construction of health care facilities and of residential housing (discussed in the next two chapters), expenditure on school construction is expected to be closely related to current expenditure in the sector, and therefore to be sensitive to the composition of the population. This is the rationale for including these three types of construction in the total expenditure in this study, rather than limiting attention to current consumption. Other types of construction—industrial and commercial buildings, transport facilities, etc.—could be estimated only by the full use of the INFORUM input-output table, as in Ridker and Watson's

Table 3-11. Public Enrollment Shares (Percentages), Historical and
Projected, by Education Level

Year	Elementary	Secondary	Higher
1930	90.2	92.8	48.4
1940	89.7	93.5	53.5
1950	87.7	89.5	51.0
1960	85.6	89.1	57.0
1970	88.6	90.9	71.6
1975	89.8	92.0	75.6
1985	90.2	93.6	80.4
2000	90.2	93.6	80.4
2010	90.2	93.6	80.4
2025	90.2	93.6	80.4

study. These three types, however, can be assumed not to depend on the
full matrix of intermediate transactions, and so can be treated like final
demand sectors.

School construction can be divided into two components: replacing
stock which has depreciated, and adding to the stock to provide schools
for larger numbers of students. Letting K_t be the stock of school buildings in year t, and assuming a constant rate r of depreciation, school
construction SC_t can be written:

$$SC_t = (K_t - K_{t-1}) + rK_{t-1}$$

$$= K_t - (1-r)K_{t-1}$$

where the term $K_t - K_{t-1}$ is net additions to the stock.

Unfortunately, there appear to be no complete and consistent data
available on the school building capital in place. Data are available on
school construction expenditures from 1945 to 1975, but because of the
postwar baby boom and the virtual cessation of construction in the preced-
ing decade and a half, it is difficult to relate historical spending to
per capita GNP or demographic variables. Several experiments with dif-

ferent functional forms and different lag structures--to allow for the
fact that buildings are not always erected just when enrollment indicates
they are most needed--failed to yield a function satisfactory for forecasting. As a result, it was necessary either to make an arbitrary judgmental forecast or to design a simple algorithm based on trends in the
education sector. We decided on the latter approach.

It seems reasonable to suppose that the stock of school capital per
student should follow the same general pattern, both across levels and
through time, as current expenditures per pupil. The desired stock should
grow with income, approaching saturation, but it should presumably grow
less rapidly than current expenditures. This is in line with the tendency
for personnel, equipment and materials spending to grow relative to expenditure on structures, in many productive sectors, and it reflects the
assumption that there is more scope for technological change in educational
activities than in the buildings where such activities occur. This reason-
ing led us to assume a constant elasticity between the capital stock (total
or per student) and current expenditure (total or per student):

$$\log K_t = a \log CE_t, \text{ or } K_t = (CE_t)^a$$

Somewhat arbitrarily, a was assumed to be 0.75 (capital requirements rise
68 percent when current expenditure doubles). The depreciation rate r was
assumed to be 0.033, reflecting a 30-year life for school buildings. These
equations make current construction expenditure a function of past enrollments (which determine the current capital stock) and of current changes
in enrollments (which determine the need, if any, for additional stock).
Because the stock is already large, and fertility has been falling, construction largely reflects depreciation. This is particularly true in
scenario F; for scenario D, additions to capacity are a much larger share
of construction. Spending for construction includes any upgrading of
school buildings, which is dependent on growth in GNP per person. There
is a further income effect through enrollment rates, but it is important
only for post-secondary schooling.

Because current salary spending is extremely sensitive to income and
productivity growth, it does not seem reasonable for the capital stock to
grow proportionately to expenditure on personnel: rather, it should depend

on the number of teachers or staff, and on spending for non-salary items.
This requirement is imposed by using only the constant-wage assumption
for salaries in calculating total current expenditure CE_t. Thus a major
difference between the growth and no-growth projections is in the relative
importance of current and capital spending. If faculty/student ratios are
constant and wages do not rise, current salary spending is simply proportional to enrollment; consequently, much of construction expenditure in
that case is also simply proportional to enrollment. Construction per
student rises only because non-salary spending rises with income, and in
the constant-income scenarios, only enrollments matter in determining construction expenditure.

Expenditure Projections

The projections of educational spending generated by the model are
disaggregated in three ways: by schooling level or age group; by public
versus private schools, and by salary, current non-salary costs, and construction. In presenting the results, however, we combine all the seven
schooling levels and also combine all current expenditures. Capital spending continues to be separated, because it is added to housing and hospital
construction to form the category of total non-business construction (sector 18) analyzed in chapter 1, and the public/private distinction is maintained because private school expenditures are part of personal consumption
expenditure (sector 176) while public spending is a separate, non-PCE sector (189). The two sectors are combined in total current educational
spending in chapter 1.

Because salaries are such a large share of total spending on education, we have, as indicated previously, made three sets of projections
rather than the two followed in all other categories. These are:

       Constant income, constant salaries

       Income growth, constant salaries

       Income growth, salaries growing at the same rate as
       labor productivity

The first and last of these correspond to the scenarios used throughout
the rest of the study, and are emphasized here and in chapters 1 and 6.

The second set was computed in order to see how important is the assumption of no growth in educational labor productivity. Some calculations were also made assuming that educational spending increases proportionately to total population rather than being related to the school-age population; these results clarify the separation of scale and composition effects in demographic change.

Constant Income Scenarios

The projections based on constant labor productivity, and therefore on nearly constant income per person and constant educational salaries, are presented in tables 3-12 and 3-13. Table 3-12 shows total expenditures, and table 3-13 shows spending per capita, both in 1971 dollars. The most striking result of these calculations is the decline in per capita expenditure, and in both its current and capital components, in all three population scenarios. Even with the most rapid demographic growth, in scenario D, enrollments decline slightly relative to population; in the absence of increases in expenditures per student, total costs also decline relative to the population.

These results are consistent with the higher school enrollment rates in the younger segments of the population, since Series D has proportionately the most population in those age groups and Series F the least proportionately. The scenarios differ in that per capita spending declines montonically in Series E and F while there is a slight dip and partial recovery in Series D. The lowest point is reached around the year 2010, after which enrollments rise rapidly. The table also shows that, although the propensity to attend private schools varies somewhat among age groups, there is no difference between the behavior of total education spending and that of public education spending, implying that there are offsetting changes underway in the age structure. It should also be noted that capital spending exhibits more of a decline over time, in all three population cases, than does current accounts spending, because net additions to the building stock become less important over time. In scenarios E and F, construction falls below half the 1975 level.

If the results of the constant income scenarios are evaluated by comparing across scenarios for the year 2025, a consistent picture emerges.

Table 3-12. Projected Current, Capital and Total Education Expenditures
(Total and Public) for Constant Income Scenarios, 1975-2025
(billions of 1971 dollars)

Scenario	1975	1985	2000	2010	2025
			Current		
D Total	63.0	66.6	78.8	86.0	102.4
Public	53.9	56.3	67.1	72.9	87.4
E Total	64.7	62.7	69.4	70.5	71.5
Public	55.6	53.2	59.5	60.1	61.0
F Total	64.8	60.5	61.4	58.1	56.3
Public	55.6	51.3	52.4	49.3	47.8
		Capital	(Construction)		
D Total	4.0	4.7	4.4	5.1	6.2
Public	3.4	4.0	3.8	4.3	5.3
E Total	5.2	3.8	3.7	3.8	3.3
Public	4.4	3.2	3.2	3.2	2.8
F Total	5.2	3.4	3.1	2.8	2.7
Public	4.5	2.9	2.7	2.4	2.3
	Total	(Current	plus Capital)		
D Total	67.0	71.3	83.2	91.1	108.6
Public	57.3	60.3	70.9	77.2	92.7
E Total	69.9	66.5	73.1	74.3	74.8
Public	60.0	56.4	62.7	63.3	63.8
F Total	70.0	63.9	64.5	60.9	59.0
Public	60.1	54.2	55.1	51.7	50.1

Series D has the highest per capita expenditures, $296, Series E is in the
middle with $246, and Series F has the lowest, $223. There is no difference among the scenarios in the public share of education expenditures,
85 percent. There is more difference among the scenarios in capital spending than there is in current accounts spending, since current spending is
related only to current enrollment levels while capital spending is partly
related to changes in those levels.

Table 3-13. Projected Per Capita Current, Capital and Total Education
Expenditures (Total and Public) for Constant Income Scenarios, 1975-2025, 1971 Dollars

Scenario		1975	1985	2000	2010	2025
				Current		
D	Total	293	273	276	270	279
	Public	250	231	235	229	238
E	Total	302	266	262	250	235
	Public	260	226	225	213	201
F	Total	304	262	245	224	213
	Public	261	222	209	190	180
				Capital (Construction)		
D	Total	19	19	15	16	17
	Public	16	16	13	14	14
E	Total	24	16	14	13	11
	Public	21	14	12	11	9
F	Total	24	15	12	11	10
	Public	21	13	11	9	9
				Total (Current Plus Capital)		
D	Total	312	292	291	286	296
	Public	266	247	248	243	252
E	Total	326	282	276	263	246
	Public	281	240	237	224	210
F	Total	328	277	257	235	223
	Public	282	235	220	199	189

Having examined the effects of age composition on educational expenditures by looking only at per capita spending, we can now isolate the
effect of population size. Table 3-14 shows the results of a computation
in which education expenditures are assumed to grow at exactly the same
rate as total population. The results are that in Series D expenditures
would increase 71 percent instead of 62 percent as in the constant income
scenario, which includes compositional effects. In Series E, they would

Table 3-14. Total Education Expenditures Projected in Proportion to
Population and Compared to Constant-Income Projections,
1975-2025 (billions of 1971 dollars)

		1975	1985	2000	2010	2025
D	Proportional	67.0	75.7	89.1	99.2	114.6
	Projection	67.0	71.3	83.2	91.1	108.6
E	Proportional	69.9	76.9	86.7	92.3	99.3
	Projection	69.9	66.5	73.1	74.3	79.8
F	Proportional	70.0	75.6	81.9	85.4	86.8
	Projection	70.0	63.9	64.5	60.9	59.0

rise 42 percent rather than 7 percent, and in Series F they would increase
by 24 percent instead of declining by 16 percent. Looking across scenarios in 2025, the range in expenditure forecasts from D to F is much greater in the constant income scenarios than it is in the forecasts scaled
simply to population: 84 percent difference versus 32 percent difference.
This indicates that the changes in age composition are reinforcing the
effects of population size, since differences in size are due almost entirely to differences in numbers of children, at least in the early years
of the projection period. Series D shows the least change in the share
of students in the total population, so the compositional effect is smallest.

Demographic elasticities (of expenditure with respect to total population) were calculated through time and across scenarios, and these are
shown in table 3-15. Across time, the elasticities are all less than 1.0
except for Series D between 2010 and 2025, when there is a sharp expansion
in the school-age population. In all periods, Series D exhibits the highest elasticity and Series F the lowest. In fact, elasticities over most
periods are negative in Series F. Looking across scenarios, the elasticities are all close to 2.0, because differences in population are predominantly differences in numbers of children, and children form less
than half the population.

Table 3-15. Elasticities of Education Spending with Respect to Population
Growth, Through Time and Across Scenarios, 1975-2025

Through time	1975-85	1985-2000	2000-2010	2010-2025
D	0.499	0.971	0.850	1.219
E	-0.514	0.823	0.254	0.958
F	-1.154	0.114	-1.694	-1.492
		1975-2000	1975-2025	
D		0.763	0.903	
E		0.211	0.378	
F		-0.499	-0.790	
Across scenarios (compared to E)	1985	2000	2010	2025
D	2.030	1.653	1.688	1.619
F	1.943	2.345	2.377	2.204

Income Growth Scenarios

The combined effect of income growth and demographic change can be
seen in the projections in which per capita income is assumed to increase
because of productivity growth. It is worth remembering here that income
growth and demographic change are not entirely independent of each other.
Due to the effect that different birth rates have on women's labor participation rates, on the relative size of the dependent population, and on
the proportion of lower-productivity younger workers in the labor force,
Series F experiences the highest growth in per capita income and Series
D the lowest. Thus we would expect income growth to offset rather than
to reinforce the effects of demographic change (as measured by the size
and relative youthfulness of the population). The model allows economic
growth to affect the projections in two ways: it partly determines the
college and graduate enrollment rates, and it is the principal determinant
of per pupil expenditures at all levels of education.

The projections of total expenditure and its current and capital
components are presented in table 3-16. Expenditures are reduced to per
capita amounts in table 3-17. Both tables show the effects first of
assuming constant educational wage or salaries, and second of assuming
that such wages rise at the same rate as overall labor productivity. The

Table 3-16. Projected Current, Capital and Total Education Expenditures
(Total and Public) for Income Growth Scenarios, 1975-2025,
with Wages Constant or Rising (billions of 1971 dollars)

Scenario		Wage	1975	1985	2000	2010	2025
				Current			
D	Total	Constant	62.6	81.7	129.9	171.4	243.6
		Rising	64.6	93.1	172.1	244.7	394.4
	Public	Constant	53.6	70.1	110.8	144.2	204.3
		Rising	55.3	79.9	147.1	206.8	333.2
E	Total	Constant	64.0	80.2	119.4	149.0	182.6
		Rising	66.0	107.3	183.2	243.4	335.5
	Public	Constant	54.6	68.6	101.6	124.8	151.7
		Rising	56.4	91.9	156.7	205.5	282.1
F	Total	Constant	64.2	79.7	110.6	129.0	149.1
		Rising	66.3	106.0	167.4	207.0	269.5
	Public	Constant	54.8	68.1	93.6	107.5	123.3
		Rising	56.6	90.8	142.6	174.2	225.9
				Capital (Construction)			
D	Total	Constant	4.0	8.9	15.0	18.9	24.7
		Rising	4.0	8.9	15.0	18.7	24.5
	Public	Constant	3.4	7.7	12.8	15.9	20.7
		Rising	3.4	7.7	12.8	15.9	20.6
E	Total	Constant	5.4	9.0	13.8	16.7	17.1
		Rising	5.4	9.0	13.8	16.5	16.9
	Public	Constant	4.6	7.7	11.8	13.9	14.2
		Rising	4.6	7.7	11.8	13.9	14.2
F	Total	Constant	5.9	9.3	12.7	14.5	14.2
		Rising	5.9	9.2	12.7	14.4	14.1
	Public	Constant	5.0	7.9	10.8	12.1	11.8
		Rising	5.0	7.9	10.8	12.1	11.8
				Total (Current plus Capital)			
D	Total	Constant	66.6	90.6	144.9	190.3	268.3
		Rising	68.6	102.0	187.1	263.4	418.9
	Public	Constant	57.0	77.8	123.6	160.1	225.0
		Rising	58.7	87.6	159.9	222.7	353.8
E	Total	Constant	69.4	89.2	133.2	165.7	199.7
		Rising	71.4	116.3	197.0	259.9	352.4
	Public	Constant	59.2	76.3	113.4	138.7	165.9
		Rising	61.0	99.6	168.5	219.4	296.3
F	Total	Constant	70.1	89.0	123.3	143.5	163.3
		Rising	71.7	115.2	180.1	221.4	283.6
		Constant	59.8	76.0	104.4	119.6	135.1
		Rising	61.0	98.7	153.4	186.3	237.7

Table 3-17. Projected Per Capita Current, Capital and Total Education
Expenditures (Total and Public) for Income Growth Scenarios,
1975-2025 with Wages Constant or Rising (billions of 1971
dollars)

Scenario		Wage	1975	1985	2000	2010	2025
				Current			
D	Total	Constant	291	335	454	539	663
		Rising	300	382	601	770	1073
	Public	Constant	249	287	387	453	556
		Rising	257	327	514	650	907
E	Total	Constant	299	340	452	528	601
		Rising	308	455	694	863	1104
	Public	Constant	255	291	384	443	499
		Rising	263	390	592	729	928
F	Total	Constant	301	345	441	497	563
		Rising	311	459	667	798	1018
	Public	Constant	257	295	373	415	465
		Rising	265	393	568	672	852
				Capital (Construction)			
D	Total	Constant	19	36	52	59	67
		Rising	19	36	52	58	66
	Public	Constant	16	32	45	50	56
		Rising	16	32	45	50	56
E	Total	Constant	25	38	52	59	56
		Rising	25	38	52	58	55
	Public	Constant	22	33	45	49	47
		Rising	22	33	45	49	47
F	Total	Constant	28	40	51	56	54
		Rising	28	40	51	56	54
	Public	Constant	23	34	43	47	45
		Rising	23	34	43	47	45
				Total (Current plus Capital)			
D	Total	Constant	309	371	507	598	730
		Rising	318	418	655	828	1140
	Public	Constant	265	319	432	503	612
		Rising	273	359	559	700	962
E	Total	Constant	324	378	504	588	657
		Rising	333	493	745	922	1159
		Constant	277	324	429	492	546
		Rising	285	423	637	778	975
F	Total	Constant	329	385	492	553	616
		Rising	337	498	719	853	1070
	Public	Constant	280	329	416	461	510
		Rising	286	427	611	718	897

assumption about compensation for educational personnel is, of course, most important for determining current spending, but it has--at least toward the end of the period--a slight effect on construction, because it affects enrollments and thereby influences capital stock requirements. A very small part of increased current spending is offset by reduced construction needs, in the future. For the reasons discussed earlier, the assumption that the wages of educational personnel rise is considered more reasonable than the assumption of constancy. The difference between the two assumptions eventually becomes very large--56 percent for total spending in the year 2025 under scenario D, 76 per ent in scenario E and 74 percent in scenario F.

Over the entire projection period, total educational spending shows rapid growth in all three scenarios, by a factor of 4.0 for Series F, 4.9 for Series E and 6.1 for Series D. The spread among the scenarios, from D to F, is considerably dampened compared to the constant income scenarios. This may be attributed partly to the offsetting effect of income mentioned above and partly to the strong effect of income growth in these three scenarios tending to overpower the differences caused by demographic change. This may also be seen in the per capita expenditure forecasts shown in table 3-17, which rise from about $300 to about $1100. The difference between Scenarios D and F in 2025 is only seven percent compared to 33 percent in the constant income scenarios. It is clear therefore that there is little net effect of population composition on total spending.

As is to be expected, capital expenditures differ relatively more among scenarios than current expenditures: they expand six-fold in scenario D, but do not even triple in scenario F, because the required stock of buildings expands much more in one scenario than in the other. In consequence, the importance of construction relative to total spending declines through time in the slower-population-growth scenarios (E and F), with most of the decline in the period 2000-2025. In scenario D the share of construction spending reaches a maximum of almost nine percent in 1985, and then returns to the 1975 level of six percent by the end of the period.

The public share of educational spending is projected to be nearly constant at about 85 percent, both through time and across scenarios.

There is a slight tendency for the share to fall after 1985, as more of
total enrollment is in higher education, where private schools have a
higher share of students. This result depends also on the cost differential, estimated at 35 percent, between private and public higher education. If costs were equal, as they are at lower schooling levels, the
public share would rise to about 88 percent in 1975 and to 86 rather than
84 percent in 2025.

Decomposition of Expenditure Growth

Total current expenditure on education (excluding construction) is
one of the twelve categories analyzed in chapter 1, and its growth through
time is decomposed into effects due to scale or population size, composition and income changes. The same analysis appears in chapter 6, for the
private component of current spending, which is a sector (176) of personal
consumption expenditure. Since the private share of total current spending is nearly constant, the analysis yields nearly identical results.
Here, we decompose the growth in total educational spending, including
school construction together with current expenditure, both public and
private. The results are given in the first part of table 3-18, for all
three scenarios and for the entire period 1975-2025 and five sub-periods.

The composition effect, reflecting the relative decline of the school-
age population, is always negative except in the period 2000-2025 in scenario D, when there is a slight relative expansion in the number of children. The slower is population growth, the more important is the compositional effect. In scenario D, the scale effect always outweighs a negative effect of changing composition, with the result that the net demographic effect is small, and income growth always accounts for about 90
percent of expenditure growth. In scenarios E and F, the scale effect is
too small to outweigh the reduction due to composition, so the net demographic effect is negative, and income growth alone would raise expenditure by the full amount projected, or even by more. Through time, in all
three scenarios, the demographic effects dwindle and the income effect
converges toward 100 percent.

Table 3-18. Scale, Composition and Income Shares of Growth in Total,
Educational Expenditure, 1975-2025, and Intermediate Periods

Scenario	Effect	1975-1985	1975-2000	1975-2025	1985-2000	1985-2025	2000-2025
		1. With Rising Wages					
D	SCA	27.3	19.0	13.5	10.6	10.4	7.1
	COM	-19.2	-6.7	-1.6	-0.3	0.3	0.5
	INC	91.9	87.7	88.2	89.7	89.3	92.4
E	SCA	15.3	13.0	10.4	6.2	6.7	3.9
	COM	-22.7	-10.5	-8.7	-1.2	-3.8	-3.3
	INC	107.3	97.5	98.3	94.9	97.1	99.4
F	SCA	12.7	11.1	7.9	4.7	4.3	1.7
	COM	-26.2	-16.1	-13.1	-4.2	-6.5	-4.2
	INC	113.5	105.0	105.1	99.5	102.2	102.5
		2. With Constant Wages					
D	SCA	37.7	28.2	23.5	16.7	18.3	12.8
	COM	-19.5	-7.4	-2.9	-0.5	0.6	0.9
	INC	81.8	79.2	79.3	83.8	81.1	86.3
E	SCA	36.9	26.1	22.6	12.2	14.4	8.6
	COM	-54.5	-21.0	-18.8	-2.3	-8.2	-7.3
	INC	117.6	94.9	96.2	90.1	93.8	98.7
F	SCA	30.3	23.0	18.1	9.2	9.5	3.7
	COM	-62.4	-33.3	-29.9	-8.2	-14.4	-9.3
	INC	132.1	110.3	111.8	99.0	104.9	105.6

It might be thought that the predominant effect of income growth is simply a consequence of assuming that educational salaries rise at the same rate as overall labor productivity. We therefore show, in the second part of table 3-18, a decomposition analysis of the projections (from table 3-16) which assume that wages are constant at 1971 levels. The effect is to reduce the income effect slightly and systematically in scenario D and--after 1985--in scenario E. However, the change is quite small; typically, the scale effect becomes larger by somewhat more than the increased compositional effect, so the net demographic component expands a little. In scenario F, however, it is the compositional effect which is enhanced more, so that income actually becomes relatively _more_ powerful in raising

Table 3-19. Income Elasticity of Total Education Expenditure, 1975-2025,
by Scenario and Period

Scenario	1975 to:	1985	2000	2025
		With Rising Wages		
D		1.516	1.331	1.408
E		2.096	1.477	1.504
F		1.793	1.408	1.423
		With Constant Wages		
D		1.014	0.911	0.943
E		1.101	0.894	0.953
F		1.008	0.888	0.923

expenditures. The impact of income on enrollment rates and on non-salary
costs is enough to account for most or all of the (considerably slower)
growth in expenditure that results in these projections.

This relation of expenditure to income emerges even more clearly when
we calculate income elasticities for the three intervals beginning in
1975, for each scenario (table 3-19). With unit personnel costs or salaries held constant, the elasticity is initially close to 1.0 and then
over larger periods declines to values between 0.89 and 0.95. This decline results from changes in enrollment rates, teacher-student ratios
and the share of total costs due to salaries. If instead salaries are
assumed to grow with productivity, the income elasticity is about 1.4 or
1.5 over the entire period, and may be a high as 2.1 in the early years.
Clearly, it is the growth in unit costs which makes education appear to be
a distinct luxury good; at constant prices, it shows an almost unitary
elasticity with respect to income. This is simply a consequence of (near)-
saturation enrollment rates at most schooling levels, and relatively little
change through time in the share of population of school age. A slight
decline in that share is almost exactly offset by increased enrollments
at higher schooling levels, making education a nearly constant share of
total disposable income.

Chapter 4

HEALTH CARE EXPENDITURES AND HOSPITAL CONSTRUCTION

The problem with regard to health care was to project total spending
on personal health care (in real dollars), whether paid for directly by
consumers or otherwise. In view of the sensitivity of health care spending to public health care programs and to other third-party payment arrangements, including private insurance, it is important to note that our
model is based on the continuation of the public policies and insurance
coverage in effect in 1975. That is, it is a relatively conservative
assumption, given recent trends in health care programs. Although this
assumption will make our projections inaccurate should some form of national health insurance be adopted, it has the advantage of not obscuring
the effects of economic and demographic changes by institutional changes,
and of avoiding the very difficult question of assessing the impact of
national health insurance on health care spending.

The model used in this study consists of age-specific utilization
rates (in dollars per capita) for both free and nonfree medical care,
which were obtained from a sample survey conducted in 1970 by the National
Opinion Research Corporation (Andersen, Lion, and Anderson, 1976). These
utilization rates change over time, being sensitive both to family size
and--via a GNP elasticity of health spending--to income. Multiplication
of the per capita rates by population in the different groups yields projected total current health spending. This model was adopted after difficulties were encountered in attempting to build a purely econometric model
from either cross-section data or time series data. Some of these difficulties are described in the next section.

Background: Relation of Health Expenditures to Income Growth

There are numerous difficulties inherent in modeling health care
spending, particularly for forecasting. These arise from the complexities
of the supply and demand relationships, which are unlike those in other
sectors of the economy. In health care, the providers are also the primary decision-makers concerning the amount of care "needed." Something
like this occurs with education also, but there the "need" is relatively
fixed, at least up to college age. Consumers are partially insulated from
the cost of care by third-party payment arrangements. Exogenous factors
such as medical discoveries are also important determinants of what can
be supplied. (This problem of technological change occurs in all sectors,
but is particularly important for health.) A further obstacle is the
dearth of statistical series on health care spending.

The Consumer Expenditure Survey reports only on consumers' out-of-
pocket expenditures for health care, which represent about one-third of
the total. Although the survey includes questions about health care received without payment, the responses to these questions are not regarded
as accurate. For these reasons the CES could not be used to project spending.

There have been cross-sectional surveys conducted specifically to
measure health care expenditures, notably by the National Opinion Research
Corporation. However, there is a fundamental obstacle to building a forecasting model by performing statistical analyses of these surveys. That
is that the behavior of health care consumption with respect to income is
very nearly opposite in the cross-section to what it is over time. In
the cross-section, the ratio of health care consumption to family income
(the budget share) decreases with increasing income (Andersen, Lion, and
Anderson, 1976). Analysis of physician utilization by members of different permanent income groups has yielded an income elasticity of only 0.27
(Andersen, Kravits, Anderson, 1975). On the other hand, in time series
data, the ratio of health care expenditure to GNP and to disposable income
has been steadily increasing as the latter has increased (Gibson and
Mueller, 1977), even after accounting for demographic change. (The adjustment of time series data for demographic change will be described

later.) Several underlying explanations can be advanced for this discrepancy. In the case of the cross-section, for instance, there is some empirical evidence (Andersen, Kravits, Anderson, 1975) that more affluent persons enjoy better health status than poorer persons, due probably to better nutrition and health knowledge, and consequently have relatively less need for corrective medical care. A second possible explanation is that there is an inverse relationship between the value of a person's time (to himself and his employer) as reflected in salary level, and the propensity to obtain medical care, a time-consuming activity. Finally, the Medicare and Medicaid programs have increased the access of low-income persons to medical care and thus reduced the importance of income as an enabling factor. It is worth noting that the low income elasticity cannot be explained by the preponderance of elderly people in the lower income groups. Among people aged 18 to 54, hospital admission rates for the lowest income group exceed those for the highest income group (Andersen, Lion, Anderson, 1976). This is not to deny that when ill, high-income persons can and typically will obtain more medical services than comparably ill poor persons.

In the case of the behavior over time of society as a whole, at least four reasons can be suggested for the increasing share of GNP that is devoted to health care. The first reason is that as society becomes more affluent it develops higher expectations about personal health and well-being as well as becoming better informed as to the benefits of medical care. The second reason is that over the years, medical research has developed ways to diagnose, treat, and perhaps cure previously unrecognized or untreatable ailments, as well as ways to prolong life even in the presence of terminal illness. This has made it worthwhile for people to obtain medical care for ailments which in the past they would have endured resignedly or unknowingly. It has also prolonged the period of treatment of the terminally ill. The third reason for the high GNP elasticity of health care is that for some time the thrust of medical research has been toward the development of rather costly techniques, such as new surgical procedures, life-sustaining apparatus and diagnostic tools which demand heavy capital outlays as well as being physician-intensive. It has been found empirically that the growth in per capita utilization of physician services between 1948 and 1966 is closely related to the changing nature of medical technology (Fuchs and Kramer, 1972). A fourth possible contribution is the increased monopoly rent of the medical profession, obtained through limitations on the supply of personnel.

Now, it is true that the second and third reasons described above are not directly income-related: they are related only to time, with which income is highly correlated. However, it can be argued that in the second case, support of medical research is partly dependent upon economic growth, that it is a luxury good. In the third case, it is plausible that if medical researchers did not believe that society would find some way to pay for the expensive new therapies, they would have pursued other lines of investigation (although this argument does not say that a different research emphasis could have reduced medical costs).

Time-series data are capable of capturing the trends just described, although not of separating income effects from time trends clearly. Unfortunately, the series of statistics on age-specific health care expenditures published by the Social Security Administration is the only one available and it begins in 1966 (see Gibson, Mueller and Fisher, 1977, and previous years). This does not provide us with enough observations for statistical analysis.

Previous Forecasting Models

As in chapter 3, we begin by reviewing a few models of health care spending which have been built by other researchers for forecasting. None of them was found suitable for this project, for the reasons outlined below.

Health Care Financing Administration Model (Freeland, Greengart and Katzoff, 1977). The HCFA model projects the value of expenditures on a particular type of health care in a future year depending upon the expenditures in the previous year, population growth, per capita utilization rate, price increase and a residual term for quality increase.

This structure has two serious defects from our point of view. First, there is no demographic detail, and thus no composition effect. The model takes account only of total population growth. Second, the per capita utilization rates are projected to continue growing at historical rates, with no allowance for the possibility of saturation, and are not linked to any economic determinants.

Trapnell National Health Insurance Model (Trapnell, 1976). The Trapnell model was designed to analyze the effect on health care spending, in 1980, of six different national health insurance proposals as well as to make a baseline projection assuming no new programs. The baseline projection was obtained separately for each type of medical care. For hospital care it was obtained by multiplying the demand-weighted U.S. population (weighted according to fixed age-specific demands) by fixed supply terms (per capita) and fixed cost terms for labor and other inputs.

To estimate use of physicians' services, Trapnell took a constant number of visits per capita times a term representing the trend toward increasing physician specialization (which leads to increased fees) and increasing services (laboratory tests, etc.) per visit. This model captures some of the demographic sensitivity we want, and some of the time trends with which we are concerned; but too many of the coefficients are constant, and those that change are not related to economic growth. This, then, is a short-run policy model rather than a long-run forecasting model.

Population Commission Model (Appleman, 1972). This model is based on data from the sample survey on health care conducted in 1963 by the National Opinion Research Corporation. Physician visits, dentist visits and hospital days were regressed separately against family size, family income, location (metropolitan area or not), and insurance coverage (yes or not), for families in three age-of-head groups. The variance explained (R^2) ranged from 0.017 for physician visits in the oldest age group to 0.75 for dental visits by families in the youngest group. This model has the advantage of including both demographic and economic variables. It has the disadvantage of using cross-sectional data to predict behavior over time, which is subject to the criticism discussed earlier. Finally, by focusing on the physical dimensions of visits and hospital days this model ignores the trend toward increasing quality and quantity of services per visit or per day which is making a substantial contribution to increasing health care expenditures. The forecasts produced by this model imply that the demand for health care will grow only 70 percent as rapidly as

GNP, which is inconsistent with the historical trend of an increasing constant-dollar share of GNP.

Use of Health Services and Current Expenditures

Data: The NORC 1970 Survey

The 1970 NORC survey of health care is the result of interviews of nearly 4000 families in early 1971 about their use of health services and the cost of those services. These responses were verified whenever possible with information from physicians, hospitals, insurers and employers.

The published results of the NORC survey (Andersen, Lion, Anderson, 1976) include data on health care expenditures for the following six individual age groups: 0-5, 6-17, 18-34, 35-54, 55-64, 65 and over, as well as information on childbirth expenses.

The survey reports on expenditures for nine types of current health care services, which we have combined into the following four categories:

    Physicians (175)
    Drugs (66)
    Tests, Eyeglasses and other Goods & Services (143 and 144)
    All Other: Hospitals, Dentists, Non-MD Practitioners (175)

These correspond as indicated to four INFORUM model sectors, but the correspondence is not one-to-one.

Although in general the survey results are not reported by family size, a special analysis was made of the effect of family size on physician utilization by adults and children (Kasper, 1975). For adults the mean number of physician visits per year was found to vary from 5.4 for members of large families (6 or more) to 6.6 for members of small families (1 to 3). This is a difference of only 22 percent between the extremes and was judged too small to be incorporated in our model.

For children, the mean number of physician visits by size of family varies more widely, as shown below:

Family size	Visits Per Year, Per Child
Small (1-3)	5.0
Medium (4-5)	3.9
Large (6 or more)	3.4

This is a difference of 47 percent from large to small families. It is particularly important to take this difference into account because the number of children per household varies among population scenarios much more than the number of adults. We have used this information to weight the per capita expenditures on physician services for children, by size of family.

A final adjustment was made to the survey's per capita expenditures to remove childbirth expenses from the expenditure rates and make them a separate new category. This was done so that childbirth expenses could be linked directly to the birth rates in the D, E, and F population projections. The procedure followed was first to compute total childbirth expenses for the U.S. in 1970, by multiplying the number of live births by the mean expenditure per live birth from the survey. We then decided that all births would be assigned to women in the 18-34 age group. (Births to younger or older women are not assigned proportionally to other age groups.) Per capita childbirth expenditure was computed by dividing total childbirth expenses by the number of persons aged 18-34. This amount was subtracted from the per capita health care expenditures given in the survey for that age group. Mean childbirth expenditures were then related directly to the number of births per year. The per capita expenditure rates for free and nonfree services, after adjustment, are shown in table 4-1.

Calibration to 1975

It was necessary to update the model to our base year of 1975 in order to take account of changes in expenditure rates which occurred between 1970 and 1975. This was accomplished by multiplying all of the age-specific expenditure rates in table 4-1 by a constant such that when the expenditure rates were applied to the numbers of persons in each age-group in 1975, the total would be equal to the estimate of total national expenditures for personal health care made by the Social Security Administration (Gibson and Mueller, 1977). The survey expenditures were inflated to 1971 dollars and the SSA 1975 expenditures were deflated using the Consumer Price Index for medical care. The updated 1975 per capita expenditure rates are shown in table 4-2. Of course, this method does not

Table 4-1. Per Capita Expenditures for Free and Nonfree Health Care by
Age, Family Size and Type of Service, 1970 (1970 Dollars
Per Year)

| Patient's Age | Family Size | Per Capita Expenditures by Service | | | | |
		Doctors	Drugs	Tests & Glasses	All Other	Total
Childbirth*	All	199.00	17.00	5.00	423.00	644.00
Under 5	1-3	60.53	15.00	7.00	38.00	120.53
	4-5	47.21	15.00	7.00	38.00	107.21
	6+	41.16	15.00	7.00	38.00	101.16
6-17	1-3	30.70	11.94	8.98	49.58	101.20
	4-5	23.94	11.94	8.98	49.58	94.44
	6+	20.88	11.94	8.98	49.58	91.38
18-34	All	54.00	25.00	18.63	105.91	203.54
35-54	All	64.00	37.91	18.96	120.00	240.87
55-64	All	97.00	56.00	34.00	192.00	379.00
Over 64	All	96.00	68.00	32.00	233.00	429.00

*This category represents expenditures per live birth and does not
double-count expenditures for children under 5 years old.

take account of any differential changes in expenditure rates among age
groups which may have occurred. To do this would have been virtually impossible given the lack of current detailed age-specific expenditure
information.

The GNP Elasticity of Health Care

In order to project the per capita expenditure rates forward, we need
to find the relationship between health care expenditures and GNP in constant dollars. The historical data for this relationship are shown below
in the first three columns of table 4-3 (Gibson and Mueller, 1977). The
data show a fairly steady increase in the percentage of GNP devoted to
health care, implying a GNP elasticity of greater than 1.0 in most years.
However, before a correct GNP elasticity can be calculated the health

Table 4-2. Per Capita Expenditures for Free and Nonfree Health Care
by Age, Family Size and Type of Service, 1975 (in 1971
dollars per year)

Patient's Age	Family Size	Per Capita Expenditures by Service				
		Doctors	Drugs	Tests & Glasses	All Other	Total
Childbirth*	All	312.03	26.66	7.84	663.26	1009.79
Under 5	1-3	94.91	23.52	10.98	59.58	188.99
	4-5	74.03	23.52	10.98	59.58	168.11
	6+	64.54	23.52	10.98	59.58	158.62
6-17	1-3	48.14	18.72	14.08	77.74	158.68
	4-5	37.54	18.72	14.08	77.74	148.08
	6+	32.74	18.72	14.08	77.74	143.28
18-34	All	84.67	39.20	29.21	166.07	319.15
35-54	All	100.35	59.44	29.73	188.16	377.68
55-64	All	152.10	87.81	53.31	301.06	594.27
Over 64	All	150.53	106.62	50.18	365.34	672.67

*This category represents expenditures per live birth and does not
double-count expenditures for children under 5 years old.

expenditures must be adjusted to remove the effects of any population age-
structure shifts which have occurred in the past. The method used was
adapted from that used by Fuchs and Kramer (1972) to estimate the effect
of demographic changes on expenditures for physicians' services between
1948 and 1966.

The age-specific expenditure rates from the 1970 NORC survey were
used to calculate hypothetical total health care expenditures for past
years, which differ from 1975 only in the number of persons in various
age-groups. These expenditures were divided by total population to obtain
per capita spending, which varies among years only in accord with differences in the age structure. The results are as follows, for six years
from 1929 to 1975:

Table 4-3. Relation of Total Health Care Expenditures to GNP in Billions
of Constant 1971 Dollars, 1929-1976, and Health Care Expenditures Adjusted to Remove Demographic Effects, Selected Years,
1929-1975

| Year | GNP | Health Spending | Health Spending as percentage | Adjusted Spending | Adjusted Share of |
	(billions of 1971 Dollars)		of GNP	(total)	GNP
1929	302.2	13.5	4.47	13.5	4.47
1935	246.4	12.3	5.00		
1940	329.9	15.2	4.61	14.4	4.36
1950	512.3	28.2	5.50	25.8	5.04
1955	628.7	33.3	5.29		
1960	707.5	42.6	6.01	39.6	5.59
1965	889.0	55.7	6.26		
1966	942.0	57.3	6.08		
1967	967.6	60.9	6.30		
1968	1010.0	64.5	6.39		
1969	1035.9	67.8	6.54		
1970	1032.5	73.4	7.11	68.5	6.64
1971	1063.4	77.2	7.26		
1972	1124.4	82.2	7.31		
1973	1185.8	85.9	7.24		
1974	1169.3	86.7	7.41		
1975	1154.3	86.4	7.49	80.0	6.93
1976	1224.0	89.1	7.28		

1929	$211	1960	$227
1940	$223	1970	$226
1950	$230	1975	$228

The historical health care expenditures from table 4-3 were then adjusted
as follows for these six checkpoint years:

$$HE_t' = HE_t \ (PCHE_{1929}/PCHE_t)$$

where HE = actual health expenditures from table 4-3

   HE' = adjusted health expenditures

 PCHE = per capita health expenditures, estimated from 1970 age-specific rates

   t = year

The resulting adjusted health care expenditures and their percent of GNP are shown, for the six years chosen, in the last two columns of table 4-3.

The disadvantage of this method is that it assumes no differential change in the expenditure rates among age groups has occurred during the period. The adjusted figures give expenditure of 80.0 billion dollars in 1975, versus actual spending of 86.4 billion; the 6.4 billion difference is attributed to changes in the age structure of the population. For 1970, the age effect raises expenditure by 5.9 billion dollars; in 1960, the effect was 4.0 billion. Over the last two decades the effect of a changing age structure has not only raised expenditure, but increased it by a rising share. Thus the extra spending attributed to demographic factors was one-tenth of the base level in 1960 but had risen to one-eighth the base by 1975.

It is now possible to calculate the GNP elasticity of health care, using the adjusted expenditures, according to the following formula:

$$\frac{(HE_1' - HE_0')/HE_0'}{(GNP_1 - GNP_0)\,GNP_0}$$

where 0 and 1 refer to the initial and terminal years of the interval for which the elasticity is calculated.

The GNP elasticities calculated are shown below in table 4-4.

Table 4-4. GNP Elasticities for Health Care Expenditures, 1929-1975

Period	Elasticity
1929–1940	0.727
1940–1950	1.432
1950–1960	1.404
1960–1970	1.589
1970–1975	1.423

Since 1940, the elasticity has been stable in the vicinity of 1.4, except for a rise during 1960-1970 which was probably due to the implementation in 1967 of the Medicare and Medicaid programs.

It seems likely that the GNP elasticity of health care expenditures will fall in the future, simply because it cannot remain so high for long without absorbing a very large share of GNP. This reason may be reinforced by some discernible trends in American attitudes toward medicine and in medical research. One factor is that the trend toward physician specialization, which tends to increase fees, may have peaked. Another is that laboratory tests are becoming increasingly automated and hence cheaper.

A more basic change is the shift in medical research toward learning about the causation and prevention of disease and toward the development and use of drugs rather than surgery or hospitalization to treat disease. A striking example of the former is the emphasis given to the links between environmental chemicals and cancer. In the latter case, an example is the impact that the development of new psychotropic drugs has had on the treatment of mental patients, many of whom are now able to avoid hospitalization. There are now indications that cancer therapy may be made more effective in the future by the development of new drugs, and perhaps surgery may be avoided in more instances than now (particularly when drugs are combined with radiotherapy). Furthermore, there are indications that in the future biochemists will be able to design drugs through molecular engineering that will have specific therapeutic actions, rather than discovering drugs by trial and error. Both of these trends would result in more cost-effective methods of dealing with disease than we have today. Lower unit costs would result in lower total spending unless the quantity of medical care demanded rose by more than enough to offset the lower cost, and that seems unlikely for currently treatable disorders. (That is, the price-elasticity of demand for treatment is probably below unity.) There will, however, be offsetting trends, to the extent that technical progress results in therapies for currently untreatable disorders.

Over the long run, it is also possible that changes in attitudes and habits, with respect particularly to exercise, diet and smoking, will result in reduced incidence of certain diseases; heart disease and lung cancer are perhaps most likely to be affected. It is still too early to

tell, however, how large a change in habits will actually occur and be
long-lasting, and what the effects on health and medical spending will
eventually be.

The future GNP elasticities which we assumed, consistent with the
trends mentioned above, are shown in table 4-5; they decline smoothly
toward 1.0 by the year 2025. Table 4-6 shows the age-specific expenditure
rates which result from applying these GNP elasticities to the intermediate
(E) population scenario for the year 2025, assuming income growth.

Public and Private Sector Allocation

Health care expenditures have been assigned to either the public sector or the private sector according to National Income Accounts rules.
That means that only direct government purchases of goods and services are
classified as public expenditures. All others, including transfer payments
such as Medicare along with private insurance payments, belong in the private sector. In the NORC survey, expenditures are given by six different
sources of payment for each age group. We have assigned those six sources
to the public or private sector as shown in table 4-7. We were then able
to compute a public/private allocation for each age group based on the
age-specific use of the sources as shown in table 4-8. These allocations
are held constant throughout the projection period, consistent with our
assumption of no program or policy changes, and no changes in the mixture
of services received by a given age group.

Hospital Construction Expenditure

Spending for hospital construction was modeled by a linear regression
of medical facilities construction against current expenditures for hospital care, using time series data published by the Social Security Administration (Social Security Administration, 1976). Other medical-related
construction, such as physicians' offices and nursing homes, was not considered. The historical data are shown in table 4-9. Additions to hospital capacity are not distinguished from replacement building. The resulting regression equation (in millions of current, not 1971, dollars) is:

Construction = 143.0 + 0.1133 (Current Spending), R^2 = 0.959

Table 4-5. Projected GNP Elasticity of Health Care Expenditures,
1975-2025

Period	Elasticity
1975-1979	1.35
1980-1984	1.30
1985-1989	1.25
1990-1994	1.20
1995-1999	1.15
2000-2004	1.10
2005-2025	1.00

Table 4-6. Projected Per Capita Expenditures for Free and Nonfree Health
Care by Age, Family Size and Type of Service, E Population
Scenario, with Income Growth, Year 2025 (in 1971 dollars)

Patient's Age	Family Size	Per Capita Expenditures by Service				
		Doctors	Drugs	Tests & Glasses	Other	Total
Childbirth	All	1481.76	126.58	37.22	3149.69	4795.25
Under 5	1-3	450.72	111.69	52.12	282.96	897.47
	4-5	351.53	111.69	52.12	282.96	798.29
	6+	306.48	111.69	52.12	282.96	753.24
6-17	1-3	228.60	88.90	66.86	369.18	753.54
	4-5	178.26	88.90	66.86	369.18	703.21
	6+	155.48	88.90	66.86	369.18	680.43
18-34	All	402.09	186.15	138.72	788.62	1515.57
35-54	All	476.54	282.28	141.18	893.51	1793.52
55-64	All	722.26	416.97	253.17	1429.64	2822.04
Over 64	All	714.81	506.33	238.27	1734.93	3194.34

Table 4-7. Assignment of Sources of Payment for Medical Care to the
Public or the Private Sector

Source of Payment	Sector
Medicaid, welfare, other free institutions	Public
Other free care (primarily military & veterans hospitals)	Public
Medicare	Private
Voluntary insurance	Private
Out-of-pocket costs	Private
Other nonfree care	Private

Table 4-8. Allocation of Health Care Expenditures to the Public or the
Private Sector by Age Group

Age Group	Percentage Share: Public	Private
Childbirth	13	87
Under 5	11	89
6-17	12	88
18-34	13	87
35-54	14	86
55-64	8	92
Over 64	9	91

Since this simple relation explains so large a share of construction expenditures, we did not experiment with more complicated models involving
lags between current and capital spending, or other explanatory factors.
At the expenditure levels of recent years, the constant term in the equation contributes little to total construction spending, which is essentially a constant fraction of current expenditures. Construction costs
are divided between public and private in the same ratio as hospital care
spending.

Table 4-9. Medical Facilities Construction Expeditures and Current
            Expenditures on Hospital Care, 1940-1970 (millions of
            current dollars)

Year	Hospital Care (current)	Facilities (construction)	Ratio Facilities/Hosp. Care
1940	$ 1,011	$ 116	0.115
1948	3,203	339	0.106
1949	3,557	660	0.186
1950	3,851	843	0.219
1951	4,254	946	0.222
1952	4,685	889	0.190
1953	5,085	686	0.135
1954	5,502	670	0.122
1955	5,900	651	0.110
1956	6,347	628	0.099
1957	6,892	879	0.128
1958	7,548	990	0.131
1959	8,177	998	0.122
1960	9,092	1,048	0.115
1961	9,921	1,174	0.118
1962	10,658	1,406	0.132
1963	11,709	1,456	0.124
1964	12,697	1,762	0.139
1965	13,605	1,912	0.141
1966	15,583	1,960	0.126
1967	18,145	2,006	0.111
1968	20,926	2,260	0.108
1969	24,093	2,973	0.123
1970	27,597	3,366	0.122

Projection Results

Table 4-10 shows projected total health expenditures, on the assumption of no growth in productivity. The scale effect is removed in table
4-11, which shows per capita expenditures on health for the constant income scenarios. The total of current and construction expenditure is also
shown, although in the twelve-category analysis of chapter 1, only current
expenditure is classified as health care; hospital construction is combined with housing and school construction. Changes over time and across
scenarios in per capita expenditures reflect only changes in the demographic composition of the population, since the effects of both population
size and economic growth have been removed. The results show slight in-

Table 4-10. Projected Current, Capital and Total Health Expenditures
(Total and Public) for Constant Income Scenarios, 1975-2025
(billions of 1971 dollars)

Scenario		1975	1985	2000	2010	2025
				Current		
D	Total	77.3	88.7	102.9	116.4	137.1
	Public	19.6	22.5	26.5	29.5	34.2
E	Total	76.7	86.6	97.5	105.5	119.8
	Public	19.4	21.9	24.9	26.5	29.3
F	Total	76.8	87.9	96.7	103.7	112.1
	Public	19.4	22.2	24.6	25.8	26.8
				Capital (Construction)		
D	Total	3.57	4.06	4.69	5.28	6.23
	Public	1.21	1.38	1.59	1.80	2.12
E	Total	3.54	3.97	4.44	4.79	5.46
	Public	1.20	1.35	1.51	1.63	1.85
F	Total	3.54	4.02	4.40	4.71	5.12
	Public	1.20	1.37	1.50	1.60	1.74
				Total		
D	Total	80.9	92.8	107.6	121.7	143.3
	Public	20.8	23.9	28.1	31.3	36.3
E	Total	80.2	90.6	101.9	110.3	125.3
	Public	20.6	23.3	26.4	28.1	31.2
F	Total	80.4	91.6	101.1	108.4	117.2
	Public	20.6	23.6	26.1	27.4	28.6

creases over time in all three scenarios, with Series F showing the greatest increase and Series D the least. Looking across scenarios, Series F
has consistently the highest expenditures and Series D the lowest, although
the difference between them is only 14 percent. Both of these findings
reflect the changing age structure of the population, with an increasing
proportion of older persons both over time and in Series F compared to

Table 4-11. Projected Current, Capital and Total Per Capita Health
Expenditures (Total and Public) for Constant Income Scenarios, 1975-2025 (1971 dollars)

Scenario		1975	1985	2000	2010	2025
				Current		
D	Total	359	364	360	366	373
	Public	91	92	93	93	93
E	Total	359	367	369	374	394
	Public	91	93	94	94	96
F	Total	360	381	386	400	423
	Public	91	96	98	99	102
			Capital (Hospital Construction)			
D	Total	16.6	16.6	16.4	16.6	17.0
	Public	5.6	5.7	5.6	5.7	5.8
E	Total	16.5	16.8	16.8	17.0	18.0
	Public	5.6	5.7	5.7	5.8	6.1
F	Total	16.6	17.4	17.6	18.2	19.3
	Public	5.6	5.9	6.0	6.2	6.6
				Total		
D	Total	376	380	376	382	390
	Public	97	98	98	98	99
E	Total	375	384	386	391	412
	Public	96	99	100	100	103
F	Total	377	398	403	418	443
	Public	97	102	104	106	108

Series D. Even in Series D, the effects of aging of the population more
than offset the effect of the relatively high birthrate and accompanying
childbirth expenses.

The components and sources of finance of health expenditures also
change over time. Capital expenditures grow very slightly slower than
current accounts spending. The declining ratio of public expenditures to
total health expenditures reflects the changing age composition of the

population, since direct government expenditures are proportionately less important for older persons than they are for the youngest age groups. (Transfer payments, however, increase; these include Medicare spending on the aged.)

Another demonstration of the effects of changing demographic composition is given in table 4-12. The constant-income projections of total health expenditure are compared to simple proportional projections where health spending is assumed to grow at the same rate as total population. This comparison shows that the divergence between the scale-only projection and the full scenario is greatest for Series F (11 percent), and least for Series D (two percent). The assumption of growth proportional to population always leads to lower estimates, since it omits the effects of the shift to an older as well as a larger population.

Table 4-13 shows yet another way of examining the interaction between demographic change and health expenditures, by computing the elasticity of health spending with respect to population, both through time and across scenarios. With one exception, in Series D between 1985 and 2000, the time-series elasticities are greater than one, ranging from 1.1 for Series D between 1975 and 1985 to 3.8 for Series F between 2010 and 2025. The highest elasticities are always found in Series F (where total growth in population is least, and aging is most rapid), and the lowest in Series D. Looking across scenarios from Series F to Series E, the elasticity ranges from -0.53 in 1975 to 0.47 in 2025. Comparing Series E to Series D, the elasticity ranges from 1.20 in 1975 to 0.81 in 2010. These findings indicate that over time the effect of aging predominates over that of mere scale, becoming very strong after 2000 in Series F. On the other hand, moving across scenarios from F to D, the increasing proportion of younger persons outweighs increases in total population size.

The next three tables show the results of the scenarios in which per capita income increases over time (tables 4-14, 4-15, and 4-16). Table 4-15 shows projected per capita health expenditures with income growth, and can be compared with the constant-productivity spending projections in table 4-11 to approximate the effect of income. (Table 4-14 shows total expenditure, and is comparable to table 4-10.) As the tables show, the impact of income growth on health expenditures is quite strong. The

Table 4-12. Total (Current and Capital) Health Expenditures Projected Proportionately to Population, 1975-2025 (billions of 1971 dollars)

	1975	1985	2000	2010	2025
Scenario Projections					
D	80.9	92.8	107.6	121.7	143.3
E	80.2	90.6	101.9	110.3	125.3
F	80.3	91.9	101.1	108.4	117.2
Proportional to Population					
D	80.9	91.3	107.5	119.7	138.3
E	80.2	88.3	99.5	105.9	113.9
F	80.3	86.8	94.0	98.0	99.6

Table 4-13. Elasticities of Health Spending with Respect to Population Growth in the Absence of Income Growth, 1975-2025, by Sub-periods

			Period	
	1975-85	1985-2000	2000-2010	2010-2025
Through time:				
D	1.107	0.926	1.163	1.148
E	1.263	1.031	1.231	1.756
F	1.758	1.165	2.108	3.763
Across		Year		
Scenarios:	1975	1985	2000	2010	2025
(Compared to E)				
D	1.20	0.70	0.68	0.81	0.69
F	-0.53	-0.71	0.15	0.20	0.47

income growth projections for the year 2025 are from 4.3 to 4.8 times as high as the constant income projections. By 1985 the growth scenario expenditures are already about 1.5 times as great as in the constant-income scenarios. This occurs despite the assumed drop in the GNP-elasticity (for a constant age distribution) from 1.35 to 1.00 over the projection interval.

Table 4-14. Projected Current, Capital and Total Health Care Expenditures
(Total and Public) for Income Growth Scenarios, 1975-2025
(billions of 1971 dollars)

Scenario		1975	1985	2000	2010	2025
				Current		
D	Total	77.3	129.2	249.2	371.2	596.7
	Public	19.6	32.9	64.1	94.1	148.9
E	Total	76.7	134.9	253.3	366.6	558.4
	Public	19.4	34.2	64.7	92.1	136.6
F	Total	76.8	142.3	266.0	379.8	542.3
	Public	19.4	36.0	67.7	94.4	130.3
			Capital (Hospital Construction)			
D	Total	3.6	5.9	11.2	16.6	26.7
	Public	1.2	2.0	3.8	5.7	9.1
E	Total	3.5	6.1	11.4	16.4	25.0
	Public	1.2	2.1	3.9	5.6	8.5
F	Total	3.5	6.4	11.9	17.0	24.4
	Public	1.2	2.2	4.1	5.8	8.3
				Total		
D	Total	80.8	135.1	260.4	387.8	623.4
	Public	20.8	34.9	67.9	99.8	158.0
E	Total	80.2	141.0	274.7	383.0	583.4
	Public	20.6	36.3	68.6	97.7	145.1
F	Total	80.3	148.7	277.9	368.8	566.7
	Public	20.6	38.2	71.8	100.2	138.6

Income growth also tends to increase the differences among the three
population series. Under constant income assumptions, the spread in per
capita expenditures in 2025 is about 14 percent from D to F. With income
growth that spread becomes 26 percent. Thus the effect of income growth
is in the same direction as the effect of demographic composition: it
leads to higher per capita expenditures in Series F than in Series D.
Much of this effect is due to the interaction between income growth and

Table 4-15. Projected Current, Capital and Total Per Capita Health Care
Expenditures (Total and Public) for Income Growth Scenarios,
1975-2025 (1971 dollars)

Scenario		1975	1985	2000	2010	2025
			Current			
D	Total	359	530	871	1167	1624
	Public	91	135	224	296	405
E	Total	359	572	958	1300	1858
	Public	91	145	245	327	450
F	Total	360	616	1061	1465	2047
	Public	91	156	270	364	492
			Capital (Hospital Construction)			
D	Total	16.6	24.1	39.2	52.2	72.7
	Public	5.6	8.2	13.3	17.8	24.7
E	Total	16.5	26.0	43.1	58.2	82.3
	Public	5.6	8.8	14.6	20.0	28.0
F	Total	16.6	27.9	47.5	65.6	92.1
	Public	5.6	9.5	16.1	22.2	31.3
			Total			
D	Total	376	554	911	1219	1696
	Public	97	143	237	314	430
E	Total	375	598	1001	1358	1919
	Public	96	154	259	346	478
F	Total	377	644	1109	1530	2139
	Public	97	165	286	386	523

demographic change described earlier: Series F shows higher rates of increase in per capita income and GNP than Series D or E, thus pushing up
health spending faster.

Income Effects and Income Elasticities

Another way to evaluate the effect of income growth is to calculate
the elasticity of projected health spending with respect to changes in

Table 4-16. Effect of Income on Total Health Spending, Projections
Based on Income Growth Alone, and Income Elasticities,
1975-2025

			Period	
	1975-85	1985-2000	2000-2010	2010-2025
Income Elasticity				
D	1.275	1.373	1.203	1.144
E	1.473	1.481	1.366	1.307
F	1.630	1.618	1.525	1.534
Income Effect as Fraction of Total Change				
D	0.781	0.828	0.732	0.707
E	0.830	0.857	0.817	0.741
F	0.832	0.885	0.831	0.810
Spending Projection Based on Income Growth but no Population Growth			Year	
(billions of 1971 dollars)	1985	2000	2010	2025
D	123.2	227.0	320.3	486.9
E	130.7	236.7	333.4	481.9
F	137.2	251.5	350.3	487.9

income. The results of such a calculation are shown in table 4-16. The
income elasticities obtained lie between 1.10 and 1.65, marking health
clearly as a luxury in the usual economic sense. The elasticities are
greatest in the early periods; this can be traced directly back to the
assumptions which were made about the relationship of health spending to
GNP over the next fifty years (table 4-5). The numbers in tables 4-16 and
4-5 differ because the latter assume a constant, 1975, age distribution,
while the former follow the changing age distribution of the population
projections. Finally, the elasticities from Series F are consistently
the highest, and those from Series D the lowest. Also, not surprisingly,
income growth accounts for the largest share of the increase in spending,
in all scenarios and periods. Income growth always accounts for between
70 and 90 percent of the increase, with all demographic effects together

Table 4-17. Decomposition of Growth in Hospital Construction Expenditure, and Income and Demographic Elasticities by Scenario and Period

Scenario	Decomposition, 1975 to:								
	1985			2000			2025		
	SCA	COM	INC	SCA	COM	INC	SCA	COM	INC
D	20.4	0.7	79.0	15.4	-0.7	85.3	10.9	0.6	88.5
E	14.1	2.7	83.2	10.6	0.8	88.5	6.9	2.0	91.1
F	10.2	6.6	83.2	7.4	2.9	89.7	4.1	3.5	92.4

Scenario	Elasticities, 1975 to:					
	1985		2000		2025	
	e_y	e_N	e_y	e_N	e_y	e_N
D	1.692	1.031	1.545	0.962	1.454	1.042
E	1.913	1.183	1.557	1.069	1.513	1.235
F	1.913	1.610	1.578	1.350	1.526	1.706

contributing only 10 to 30 percent. The income component is always largest in scenario F, where population growth is slowest.

Current health care expenditures, excluding hospital construction, form one of the twelve consumption categories analyzed in chapter 1. Health care absorbs just over nine percent of total consumption and construction in 1975. In the absence of productivity growth, the share would decline very slightly for scenario D, increase slightly for E, and rise by one percentage point for F (see table 1-6). With income growth, however, the share would approximately double in all scenarios, reaching the range of 15 to 18 percent. Income therefore clearly accounts for the largest share of total growth, as is confirmed in table 1-7: the income share is around 80 percent for the first projection decade (1975-85), and about 90 percent over the entire half-century. Composition effects are always quite small; only in scenario F do they approximate the size of the population scale effect. Income elasticities (see table 1-8) stay close to 1.5 for any projection period beginning in 1975; demographic elasticities

are about 1.1 or 1.2 in the faster-growing scenarios, but reach a value
of 1.7 for scenario F.

Since hospital construction is close to being a constant share of
hospital care expenditures, which in turn are a nearly constant fraction
of total current expenditures, construction has to show nearly the same
behavior as current spending. We nonetheless include, in table 4-17, a
decomposition analysis of hospital construction, and the corresponding
elasticities. This permits hospitals to be separated from home and school
construction, whereas in chapter 1 all three kinds of construction are
aggregated. As is evident from the table, growth in hospital construction
is due chiefly to income growth, the share rising through time and as
population growth is slower. For this reason, and because a change in
the age composition acts both to raise and to lower current health care
spending, the compositional effect is usually below three percent. Income
elasticities show the same pattern as income effects, and range in value
from 1.5 to 1.9. Population elasticities differ sharply from 1.0 only
for scenario F.

Chapter 5

HOUSING CONSTRUCTION

A model for projecting the value of residential construction in
response to changes in income and demographic composition and growth was
constructed using information from the Consumer Expenditure Survey. The
model generates projections of the number of new units to be built each
year, and their value, for single family homes, apartments and mobile
homes. Duplexes and townhouses are treated as single-family homes;
"apartments" house more than one family and have a common entrance. Projected construction in each year consists of replacement units and additional units. Repair and maintenance construction and additions to
existing dwellings are not included; such expenditures are disaggregated
among consumption sectors, classified by the used material or the service
performed. The independent variables of the model are the housing preferences of forty-two types of households, as defined by family size and
age of head, and mean household incomes for households in single-family
homes, apartments and mobile homes.

Background: Previous Analyses

There are numerous models of the housing market, many of which are
short-term models focusing only on predicting fluctuations in construction
activity. Long-range models are generally of two types. The first consists of a single econometric equation such as Almon et. al. (1974) use
in INFORUM, in which the independent variables are per capita disposable
income, interest rates, replacement rates and the total number of households, and the dependent variable is the amount of conventional construction (single family and apartments but not mobile homes) in constant

dollars. The second type is exemplified by the U.S. Forest Service model
(Marcin, 1974) which projects the number of units needed by type according
to the number and housing preference of households in various age-of-head
groups. Neither of these long range models combines the desired demographic detail with the ability to project the value of residential construction activity, including mobile homes. For this reason, and because it
was desirable to use the same data base as used for the majority of the
demographically-sensitive consumption equations, an original model was
constructed from the Consumer Expenditure Survey data.

Structure of the Model: Housing Choices and Values

The basic structure of the model consists of two parts. The first is
a matrix of housing preferences (expressed as percentage shares) shown in
table 5-1, for forty-two household types and three housing types. This
matrix was constructed from survey responses indicating the type of dwelling unit in which the household was living at the time, and is assumed to
be constant through time and independent of income. The basis for this
assumption lies in a series of linear regressions using the CES data, in
which housing preferences were regressed against income for each of the
forty-two household types. The majority of these (one-variable) regressions yielded R^2 statistics of 0.1 or less, indicating that income is not
a significant factor in determining the choice of housing type for households in a given size and age of head category. In fact, income almost
surely does have some effect on housing preference, but its influence is
difficult to capture in the cross-section, since a great many other fac-
tors--location, type of occupation or business, etc.--may also matter. A
more serious difficulty is that projections based on an income effect
might, with the growth in absolute incomes, lead to implausible results
such as the complete disappearance of apartments in favor of single-family
homes. Probably relative rather than absolute incomes matter, and preferences may change through time because of environmental changes which will
be correlated with income but not due to it. Removing all income effects
seemed the safest course. It is least defensible in the case of mobile
homes, which may be "preferred" largely because they are inexpensive. If

Table 5-1.   Matrix of Housing Choices by Age and Family Size (Percentage
Shares)

Household Size	Housing Type	Age of Head						
		15-24	25-29	30-34	35-44	45-54	55-64	65+
1	Single family	24	21	23	33	44	54	57
	Apartments	65	74	67	56	48	37	36
	Mobile homes	11	5	10	11	8	9	7
2	Single family	33	47	65	67	82	86	75
	Apartments	58	47	30	22	14	11	21
	Mobile homes	9	6	5	11	4	3	4
3	Single family	40	65	78	86	89	88	79
	Apartments	51	27	16	11	10	10	19
	Mobile homes	9	8	6	3	1	2	2
4	Single family	45	70	84	88	89	93	86
	Apartments	45	25	12	12	8	7	14
	Mobile homes	10	5	4	0	3	0	0
5	Single family	50	85	88	93	94	100	100
	Apartments	50	15	8	5	6	0	0
	Mobile homes	0	0	4	2	0	0	0
6 or more	Single family	100	86	88	93	96	100	100
	Apartments	0	14	12	6	4	0	0
	Mobile homes	0	0	0	1	0	0	0

this is so, rising income will reduce this preference, whereas the model
holds it constant.

It is essential to note that this model attempts to describe and
predict the type of dwelling in which people will live. It says nothing
about ownership versus renting, or about the purchase of homes--of what-
ever type--as investments. It does not matter that some households own
several homes while others own none; what matters is that every household
has a dwelling, and the type of dwelling is assumed, somewhat unrealistic-
ally, to be independent of the choice between owning and renting. Since

the entire projection apparatus for all expenditure sectors takes no
account of households' asset portfolios and assumes constant relative
prices (including that between housing and other sectors), we avoid such
questions as whether shifts in owning/renting relations--particularly the
expansion of condominium ownership of apartments--will affect preferences
for housing types. The housing projections deal with physical, not financial, choices. This may be the weakest assumption in the model, because
inflation has recently been changing the value of housing as an asset and
increasing the incentive to own rather than rent. It is not clear that
such a change has no effect on preferences for the physical type of housing.

The second part of the model translates numbers of units into value
of construction by means of three linear equations. In these equations,
CES data on dwelling unit market values were regressed against estimated
permanent household income, the same variable used in the interview survey
consumption equations, for households living in each of the three dwelling
types. Market value and construction cost are assumed to be equal: no
allowance is made for land rents, because the survey data do not distinguish rents and costs. The assumption of constant relative prices in the
future means that cost increases reflect increased physical requirements;
the price per "unit" of housing is constant, and as incomes rise, households buy more such units, that is, larger or better-constructed dwellings.
The projections of mean household income, which are discussed in chapter
2 and used as the independent variable in these equations, are shown in
table 5-2 for each housing type, in each population scenario, both with
and without growth in productivity. Total value of construction for each
dwelling type is the product of the projected number of units constructed
and the projected mean market value, which is derived from the mean incomes of households preferring that type of dwelling.

Components of Requirements for New Housing

The projection of housing construction requirements must take into
account several distinct components of the need for new housing: replacements, additions and second homes. Replacement requirements are defined

Table 5-2. Projected Mean Household Incomes by Population Series, Income Growth and Housing Type, 1976-2025 (1971 dollars)

Year	Series D			Series E			Series F		
	Single	Multi	Mobile	Single	Multi	Mobile	Single	Multi	Mobile
With Income Growth									
1976	8624	6096	5266	8550	6047	5225	8601	6065	5239
1980	9567	6796	5874	9721	6900	5968	9903	6963	6024
1985	9981	7128	6164	10156	7247	6276	10638	7440	6446
2000	14177	10143	8769	14553	10338	8957	15034	10520	9121
2010	16473	11844	10234	17096	12118	10504	17706	12423	10768
2025	19913	14410	12446	20531	14679	12702	21647	15220	13188
Without Income Growth									
1976	8862	6263	5411	8797	6222	5375	8722	6150	5313
1980	8571	6089	5263	8457	6003	5192	8390	5899	5104
1985	8409	6005	5193	8192	5845	5062	8169	5713	4950
2000	8229	5887	5090	7936	5637	4884	7733	5411	4692
2010	8044	5784	4998	7795	5525	4789	7518	5275	4572
2025	8146	5895	5091	7816	5588	4835	7661	5387	4668

as the product of a replacement rate and the previous year's stock. Additions are equal to the net number of new households formed since the last year. Second homes are a percentage of both additions and replacements. The equation below represents these relationships:

$$HR_t = (1+e)[HH_t - HH_{t-1} + rS_{t-1}]$$

where HR = new housing requirements (units)

S = housing stock (units)

HH = households (number)

r = annual replacement rate (0.008, 0.007 or 0.050, depending on housing type, as described below)

e = rate of ownership of second homes (0.04)

t = year

Projection of Additional Housing Units Needed

The number of additional housing units needed of each type is determined by the year-to-year change in the number of households in each age and size category (see table 5-3 for an example) and the housing preferences previously determined. Since this is a demand-driven model, possible supply constraints play no role in determining the number and type of units built. Table 5-4 shows the results of such a computation. Negative numbers in tables 5-3 and 5-4 indicate a decrease in the number of households of a specific type or in the number of dwelling units they require of each type. These are concentrated among large households, which become less common in the future as the share of children in the population declines.

Projection of Replacement Housing Units

The number of replacement housing housing units needed of each type is determined by the stock of units in the previous year and an annual replacement rate exogenously assumed for each type of housing. The replacement rates used are shown below, and a sample projection of replacement units needed is shown in table 5-5.

Table 5-3. Sample Calculation of Year-to-Year Change in Number of
Households by Age and Size (thousands)

Family size	Age of Head							
	15-24	25-29	30-34	35-44	45-54	55-64	65+	All
1	64	75	68	50	-13	43	204	491
2	100	123	59	79	106	267	314	1048
3	64	101	62	101	150	11	-44	444
4	12	51	59	97	20	-69	-88	83
5	-1	-4	-9	1	1	-2	-1	-14
6 or more	-2	-9	-30	-113	-58	-10	-2	-224
All	239	337	209	215	205	240	383	1827

Single-family home	0.008
Multi-family home	0.007
Mobile home	0.050

Projection of Second Homes Needed

Second homes were projected as a constant four percent of primary
homes required of each type. A second home is used by a single household
which already has a dwelling; the term does not refer to investment prop-
erty. There were insufficient data available in either the CES or in time
series data to estimate an equation for the demand for second homes. Undoubtedly, second-home ownership is income-elastic, but it is consistent
with our assumption that preferences are independent of income to ignore
any such effect. Second homes are not shown separately in the projections
but are included in projected numbers of additional and replacement units
needed.

Conversion from Physical to Financial Value

The three equations relating housing market value to household income, as initially estimated from the CES data, are as follows (dependent and independent variables are in 1971 dollars):

Table 5-4. Sample Calculation of Additional Housing Units Needed
by Housing Type and Age and Household Size (thousands)

Family	Age of Head							
Size	15-24	25-29	30-34	35-44	45-54	55-64	65+	All
				Single-Family				
1	13	14	14	15	-5	20	100	171
2	28	48	34	46	75	199	202	631
3	22	58	41	80	14	8	-30	294
4	5	31	44	74	15	-56	-66	47
5	-0	-3	-6	1	1	-1	-1	-10
6 or more	-1	-6	-23	-89	-47	-8	-2	-176
All	66	142	104	127	153	162	203	956
				Multi-Family				
1	36	46	39	23	-5	14	64	217
2	51	51	15	17	12	24	58	228
3	28	24	9	11	13	1	-7	78
4	5	11	6	8	2	-3	-7	20
5	-0	-0	-1	0	0	-0	-0	-1
6 or more	-0	-2	-3	-6	-2	-1	0	-14
All	119	129	65	53	19	36	106	527
				Mobile Homes				
1	6	4	6	5	-1	3	12	35
2	8	7	1	5	4	6	10	42
3	5	5	3	2	1	0	-1	16
4	1	2	1	1	0	-1	-1	4
5	0	-0	-0	0	0	-0	0	-0
6 or more	0	-0	-0	-1	-1	0	-0	-3
All	21	19	11	12	4	9	20	95
				All Types				
1	55	64	58	43	-11	37	176	423
2	86	106	51	68	91	230	270	901
3	55	87	53	93	129	9	-38	388
4	11	44	51	84	17	-59	-75	71
5	-1	-3	-7	1	1	-1	-1	-12
6 or more	-1	-8	-26	-97	-50	-9	-2	-193
All	205	290	180	191	176	206	329	1578

Table 5-5. Sample Calculation of Replacement Housing Units Needed
by Housing Type and Age and Household Size (thousands)

Family Size	Age of Head							
	15-24	25-29	30-34	35-44	45-54	55-64	65+	All
				Single-Family				
1	4	3	2	3	6	11	31	60
2	7	8	5	8	28	47	47	150
3	6	13	9	14	20	13	8	83
4	2	11	16	26	19	10	7	92
5	1	5	9	23	14	3	1	56
6 or more	0	2	6	22	10	2	0	43
All	20	42	47	96	97	87	94	483
				Multi-Family				
1	10	9	5	5	5	7	17	57
2	11	8	2	3	4	5	12	44
3	7	5	2	2	2	1	2	20
4	2	3	2	2	2	0	1	12
5	0	1	1	1	1	0	0	4
6 or more	0	0	1	1	0	0	0	3
All	29	25	12	14	14	14	32	140
				Mobile Homes				
1	12	6	5	6	6	12	23	70
2	12	8	1	5	9	10	15	60
3	10	7	4	2	1	2	1	27
4	2	6	3	3	2	1	1	18
5	0	1	2	1	1	0	0	6
6 or more	0	0	1	2	1	0	0	4
All	36	27	16	20	22	24	40	186
				All Types				
1	25	17	11	14	17	31	72	187
2	30	23	8	15	41	62	73	254
3	23	25	14	18	24	16	11	131
4	6	20	21	31	23	11	9	122
5	1	6	12	26	16	4	1	66
6 or more	0	2	8	25	12	2	1	50
All	85	95	75	130	133	125	166	809

Single-family homes:

  Market value = 10,312 + 1.373 (single-family home dweller's
                                         family income)

Apartments:

  Market value = 19,924 + 0.932 (apartment dweller's family income)

Mobile homes:

  Market value = 6,270 + 0.249 (mobile home dweller's family income)

As the equations indicate, the value of single-family homes has the
highest income coefficient and mobile homes the lowest, indicating that
single homes are found in a much wider price range than are mobile homes,
whose value is nearly constant with respect to income. However, these
results were unsatisfactory in several respects. For example, the average
value of an apartment exceeds the value of a single family home at incomes
below $21,800, although at higher incomes the positions are reversed. This
greatly overstates the housing requirements of low-income apartment dwell-
ers. The calculated value of mobile homes was also not in good agreement
with statistical reports. Finally, the equations yielded an income response of housing construction expenditures considerably lower than other
studies have shown (Reid, 1962). These problems are attributable to
(1) inaccurate estimating of market values by respondents (market values
were not known exogenously, and rental data were not used), (2) small
sample size in the case of apartments (values were reported in the survey
only for owner-occupied units), and (3) differences between behavior with
respect to income over time and in the cross-section. All these problems
might act to make the constant term too large and the income coefficient
too small (because of errors and transitory variation), and the bias
should be greater for apartments and mobile homes than for single-family
houses. The final equations, modified to correct the deficiencies noted,
are shown below.

  Single family:

    market value = 2041 + 2.318 (income)

  Apartment

    market value = 4472 + 1.784 (income)

  Mobile:

    market value = 2329 + 0.790 (income)

With this adjustment, the cross-over income for equality between apartments and single-family home values is only $4,550. As before, mobile
homes are the cheapest alternative at all income levels.

As a final step the model was adjusted by a simple multiplicative
factor of 0.8 so that the calculated number of units and market values
would be approximately equal to statistics for 1975 as reported in Construction Reports. An exact match was not necessary or desirable since
the purpose is to simulate not short-run fluctuations but long-run trends.
This adjustment was necessary because the model was found to overstate the
number of units built each year in the past; this is probably a consequence of relative price variation, so that dollar increases do not trans-
late directly into physical units.

Summary of Calculations

The structure of the moᴅᴇʟ for the housing projections may be summarized in the following steps:

(1) Projections of the numbers of households in forty-two categories
defined by size and age-of-head, described in chapter 2.

(2) Projections of mean household income for all types of households
living in each of three dwelling types, also described in chapter 2.

(3) A matrix of housing preferences for three dwelling types and
forty-two household types, from the CES data.

(4) The number of additional housing units needed of each type,
projected by applying the housing preference matrix to the year-to-year
change in the number of households by type.

(5) The number of replacement housing units needed, projected by
applying a replacement rate to the previous year's stock of housing units
of each type.

(6) The number of second home units needed, calculated as a constant
percentage of the number of primary (addition plus replacement) units
needed by type.

(7) Total housing construction requirements, defined as the sum
of additions, replacements and second homes.

(8) Equations relating the mean value of a new dwelling unit of each type to mean household income.

(9) The value of new construction was calculated by multiplying the mean value of a new unit of each type, as determined by the equations obtained in step (8), by the projected number of new units of each type.

Projection Results

The results of the projections made with the housing model are shown in tables 5-6 through 5-9. The base year is 1976 rather than 1975, because the model starts with the 1975 housing stock. Tables 5-6 and 5-7 refer to the scenarios that assume constant productivity, or nearly constant per capita income, under the three population growth assumptions, Series D, E and F: Table 5-6 shows totals, and Table 5-7 shows per capita values. Table 5-6 also shows housing construction expenditure projected to grow at the same rate as population, with no allowance for compositional ef- fects. The scale effect alone leads to a smaller projection out to 1985, but an increasingly larger one thereafter. The results for the year 2025 are quite dramatic. Based on population growth alone, Series D expendi- tures would be about $60 billion rather than $46 billion; Series E would be about $49 billion rather than $31 billion; and Series F would be $42 billion rather than $21 billion. This indicates that by and large the effect of changes in demographic composition is to reduce housing construc- tion. Tables 5-8 and 5-9 present the results of the income-growth sce- narios for the same population series. Conventional housing (single family houses and apartments) is shown in total; mobile homes are separated be- cause they correspond partly to sector 140 (trailer coaches) of personal consumption expenditure. The following discussion analyzes these results by separating the effects on residential construction expenditures of the three basic determinants: population size, demographic composition, and income growth.

Compositional Effects

When per capita expenditures are calculated from the constant income projections, as shown in table 5-7, they show the effects of changes in

Table 5-6. Projected Housing Construction Expenditures for Constant
Income Scenarios, 1976-2025 (Billions of 1971 dollars)

Scenario	1976	1985	2000	2010	2025
Conventional (Single-Family and Apartment)					
D	33.4	40.6	35.0	37.0	43.7
E	33.1	39.5	29.1	33.3	28.5
F	32.7	39.2	26.0	29.6	19.2
Mobile Homes					
D	1.4	1.9	1.9	2.1	2.5
E	1.5	2.0	1.9	2.1	2.1
F	1.5	2.0	1.8	1.9	1.9
Total					
D	34.9	42.5	36.9	39.0	46.2
E	34.6	41.5	31.0	35.4	30.6
F	34.2	41.2	27.8	31.5	21.0
Projection Proportional to Population Size Alone Total					
D	34.9	39.4	46.4	51.6	59.7
E	34.6	38.1	42.9	45.7	49.1
F	34.2	36.9	40.0	41.7	42.4

Table 5-7. Projected Per Capita Housing Construction Expenditures for
Constant Income Scenarios, 1976-2025 (in 1971 dollars)

Scenario	1976	1985	2000	2010	2025
Conventional (Single-Family and Apartment)					
D	153.37	166.46	122.38	116.28	118.91
E	153.34	167.59	110.06	118.09	93.81
F	152.21	169.77	103.71	114.15	72.48
Mobile Homes					
D	6.43	7.79	6.64	6.60	6.80
E	6.95	8.49	7.19	7.45	6.91
F	6.98	8.66	4.71	7.33	7.17
Total					
D	160.26	174.25	129.02	122.56	125.71
E	160.34	176.07	117.25	125.5	100.72
F	159.09	178.43	110.89	121.48	79.28

Table 5-8. Projected Housing Construction Expenditures for Income Growth Scenarios, 1976-2025 (billions of 1971 dollars)

Scenario	1976	1985	2000	2010	2025
Conventional (Single-Family and Apartment)					
D	32.9	46.9	56.3	69.5	97.0
E	32.5	47.0	49.1	66.0	66.5
F	32.4	48.6	46.1	61.8	47.5
Mobile Homes					
D	1.4	2.1	2.8	3.4	4.8
E	1.4	2.3	2.8	3.5	4.2
F	1.7	2.3	2.7	3.4	3.8
Total					
D	34.3	49.0	59.1	72.9	101.8
E	34.0	49.0	51.9	69.5	70.7
F	33.9	50.9	48.7	65.1	51.3

Table 5-9. Projected Per Capita Housing Construction Expenditures for Income Growth Scenarios, 1976-2025 (1971 dollars)

Scenario	1976	1985	2000	2010	2025
Conventional (Single-Family and Apartment)					
D	151.1	192.29	196.85	218.42	263.95
E	150.6	199.41	185.70	234.04	218.89
F	150.7	210.48	183.88	238.33	179.31
Mobile Homes					
D	6.4	8.61	9.79	10.69	11.43
E	6.5	9.76	10.06	12.41	13.82
F	7.9	9.96	10.77	13.11	15.86
Total					
D	157.5	200.90	206.64	227.84	277.00
E	157.6	209.16	196.29	246.45	232.72
F	157.7	220.43	194.26	251.06	193.66

the household composition (family size and age of head) of the population across time and among population series. In the constant income scenarios mean household income varies directly with household size, which tends to be greater with higher population growth rates. Household income under the constant income assumption therefore tends to decline over time. Income per person does not decline, but housing value is assumed to depend on total, not per capita, income in the household. The result of these demographic influences is that per household income decreases in all three constant income scenarios, and it decreases least in Series D and most in Series F. Thus the constant income projections of housing construction expenditures capture the indirect effects of demographic change via income, as well as the direct effects of demographic change on housing construction requirements.

Table 5-7 shows that there is a 22 percent decline in per capita construction expenditures by 2025, in Series D. This decline is not continuous over time; per capita expenditures peak in 1985, then decline, then rise slightly between 2010 and 2025. Excluding demographically induced income effects, the major explanation for the decline is a slight shift toward mobile homes, which are the least expensive form of housing construction. This shift is the result of a slightly greater preference for mobile homes among younger families (household heads aged 34 or less), which grow in numbers somewhat faster than older families—assuming, of course, that housing preferences do not change.

In Series E the decline in per capita construction spending is 37 percent. This reflects a greater demographically-induced decline in household income as well as a greater shift toward mobile homes and away from both single-family homes and apartments. It is also due to a slowdown in population growth over time. There were 1.3 million more households in Series E in 1976 than there were in 1975. Between 2024 and 2025, only 0.6 million households are added. This means that only half as many additional housing units per year are needed.

The per capita decline in construction spending in Series F is even greater—50 percent. It is also due to the combination of household income variation, shifts among housing types, and population growth slowdown. In this series the shift toward mobile homes is substantial, and perhaps exag-

gerated: from 12 percent of new units in 1976 to 22 percent in 2025. The
decline in additional housing units needed each year is also striking:
from 1.1 million in 1976 to 0.1 million in 2025.

The combined effects of income and population growth are captured in
table 5-8, which shows the results of the total construction spending projections for the income growth scenarios. The increases over time range
from about 150 percent in Series F to about 250 percent in Series D, compared to 60 percent and 132 percent for the constant income scenarios.
The spread across the income growth scenarios in 2025 is a ratio of 1.98,
while the ratio is 2.20 in the constant income cases, indicating that
income effects are slightly offsetting to population effects. This is to
be expected, since households in Series F have higher mean incomes than
Series D households in the income growth scenarios.

The per capita projections in Table 5-9 similarly demonstrate the
striking effect of income growth when compared to the constant income projections. . In Series D the income growth projections increase housing
expenditure 76 percent, compared to a decline of 22 percent in the constant income case. Series E grows by 48 percent compared to a decline of
37 percent, and Series F shows a 23 percent increase compared to a 50 percent decrease in the constant income case.

The model assumes first, that replacement rates are constant for any
given type of dwelling and second, that the share of second homes is constant. In consequence, the number of dwellings per household is constant,
and the average life of a housing unit varies only with the housing type
mix and thus only with the age and household size composition of the
population. This means that the increased expenditure associated with
income growth represents bigger and better housing but not more housing
units: more rooms per person, for example, rather than more dwellings or
a faster turnover. Both assumptions are conservative, in the sense that
income growth might in fact, aside from causing shifts among housing types,
lead to more second homes or to a shorter physical life and more rapid
replacement for dwellings. The projections are more likely to understate
than to overstate housing construction, for these reasons. Of course,
this analysis depends on another crucial assumption, that of constant
relative prices. If the relative price of housing rises in the future--

whether because of land rents or increased construction costs--then
replacement rates might fall, so that dwellings would be used longer, and
housing construction might be overstated by the model.

Decomposition of Expenditure Growth

The category-level analysis in chapter 1 treats all three kinds of
construction--housing, schools and hospitals--together. We therefore
show, in Table 5-10, the components of expenditure growth and the income
and demographic elasticities, for housing construction alone, for each
of three periods beginning in 1976. Because these parameters do not vary
monotonically, the table also shows elasticities for intermediate intervals starting in 1985, 2000 or 2010. (The decomposition analysis is not
repeated for these intervals, but can be derived from the projections in
tables 5-6 through 5-9 and the population and income projections in chapter 1.)

The most dramatic result in table 5-10 is the very large composition
effect, always over 20 percent in absolute value and exceeding 100 percent
over the entire period for scenario F. Few kinds of expenditure are nearly
so sensitive as housing construction, to shifts in the household composition of the population. For most personal consumption sectors, as is
shown in chapter 6, composition effects die out over time, but for housing
they can increase as the projection period is longer. This behavior has
two notable consequences. First, income effects have to offset the large
negative composition effects beyond 1985, and therefore can exceed 100 percent. Second, demographic elasticities (e_N) depart markedly from 1.0 and
can become negative beyond 1985. There is no convergence toward a pattern
of unitary demographic elasticities and small overall demographic effects.
Although income elasticities (e_Y) do converge toward 1.0 (or at least
toward 0.8) when measured from 1976 forward, they increase beyond unity
when measured over the intermediate periods 1985-2000 and 2000-2025. Over
these intervals, as the last part of the table shows, population elasticities tend to be negative and can be as large as -5.0, whereas the income
elasticities come to exceed 2.0, and reflect a growth in expenditure where
demographic factors alone would reduce it.

Table 5-10. Decomposition of Growth of Housing Construction Expenditure, and Income and Demographic Elasticities, 1976-2025

Scenario	Decomposition, Period 1976 to:								
	1985			2000			2025		
	SCA	COM	INC	SCA	COM	INC	SCA	COM	INC
D	32.9	21.0	46.1	47.3	-39.1	91.7	36.9	-20.0	83.1
E	24.5	23.5	52.1	47.2	-68.0	120.8	40.3	-51.3	111.1
F	16.8	25.1	58.1	41.2	-85.4	144.1	48.3	-125.5	177.2

Scenario	Elasticities, Period 1976 to:					
	1985		2000		2025	
	e_Y	e_N	e_Y	e_N	e_Y	e_N
D	0.603	1.579	0.774	0.196	0.824	0.525
E	0.646	1.876	0.768	-0.518	0.813	-0.350
F	0.643	2.358	0.769	-1.286	0.809	-2.255

Scenario	Elasticities, Intermediate Periods			
	1985-2000		2000-2025	
	e_Y	e_N	e_Y	e_N
D	1.264	-0.889	2.258	0.896
E	1.274	-2.536	2.330	-0.094
F	1.399	-4.788	2.389	-5.086

These results show so much variation because they refer to <u>changes</u> in the housing stock, which are naturally more volatile than the stock itself or than most consumption flows. This explanation applies equally to school and hospital construction, discussed in chapters 3 and 4. The additional volatility of housing construction is due to the fact that expenditures are determined directly by the <u>household</u> composition of the population, and this is more variable than the age-group composition, which is all that matters for health care and education. The introduction of three housing types then adds a little more variation: no such distinction is

made for construction of schools by level or of hospital facilities by
the age of the patients who use them. The difference between the growth
of housing construction and the growth of housing services can be seen by
comparing the decomposition and elasticity analysis of table 5-10 with the
corresponding results for sector 167, value of imputed rent to owner-
occupied dwellings, taken from tables 6-10 through 6-12. The latter show
both income and demographic elasticities always close to 1.0, and an income
component of growth which always lies between 71 and 91 percent. Composition effects, which are so important in construction, are negligible determinants of the growth of housing services:

1975 to:	1985			2000			2025		
Scenario	SCA	COM	INC	SCA	COM	INC	SCA	COM	INC
D	27.7	1.0	71.4	21.8	0.4	77.8	18.3	0.2	81.6
E	22.4	-0.6	78.2	17.2	-0.2	83.1	13.3	-0.1	86.8
F	18.1	-1.2	83.1	13.4	-0.5	87.1	8.8	-0.2	91.4

	e_Y	e_N	e_Y	e_N	e_Y	e_N
D	1.176	1.032	1.034	1.014	1.090	1.008
E	1.053	0.975	0.977	0.988	1.043	0.993
F	0.915	0.936	0.933	0.969	1.003	0.977

Chapter 6

PROJECTIONS OF PERSONAL CONSUMPTION EXPENDITURE BY SECTOR

Expenditure was separately projected for each of the 133 sectors comprising personal consumption expenditure, which together account for nearly 90 percent of total expenditure considered. Six of these sectors (66, 140, 143, 144, 175 and 176) are related to health, education, and housing and were projected as described in chapters 3 through 5. Another 76 were estimated entirely from the consumer expenditure survey, by the procedure discussed in chapter 2; for the remaining 51 sectors, we used Ridker and Watson's aggregate, time-series projections. Results for all the PCE sectors are presented in this chapter: this analysis disaggregates the results shown in chapter 1, where demand was divided into only twelve categories, eleven of which consist partly or entirely of components of PCE. The results for sector 18, new construction of homes, schools and hospitals, appear in total in chapter 1 and separately in chapters 3 through 5, and so are not repeated here.

As we noted earlier, projections based on the CES data would sometimes lead to unreasonable expenditure levels, because the cross-section data take no account of eventual saturation or other departures from the income-expenditure relation in the short run. A further difficulty is that the CES data may under- or over-estimate expenditure in the base-year, households' responses being rather unreliable for some classes of expenditure. Before presenting the projection results, therefore, we discuss the adjustments made to compensate for these two problems.

Adjustment of Demand Equations for Projection

Of the 76 sectors estimated from the survey data, 25 required some adjustment in order to be used for projection, because a simple projection on the basis of income growth would lead to what we considered unreasonable total expenditure in the sector. "Unreasonable" is, of course, a matter of judgement, and is to some degree arbitrary; the model is not complete enough to exclude such judgements. Typically, the cross-section equations would outrun acceptable exogenous projections of demand, but in some cases the cross-section estimate would be, in our opinion, too low.

The simplest adjustment is to set an income coefficient equal to zero if the regression estimate was negative for one or more size classes. This prevents negative expenditure projections, forcing absolute saturation and independence of income growth. This adjustment was applied to sectors 34 (tobacco), 35 (fabrics), 80 (pottery), 140 (trailer coaches), 148 (toys), 152 (buses), 158 (telephone and telegraph), and 169 (hotels and lodging places). Direct adjustments were also made to income coefficients for sectors 35 and 124 (lighting and wiring) when regressions for one or more household size classes deviated markedly and inexplicably from the rest.

Expenditure in sector 52 (newspapers) was held constant in each size class, suppressing all effects except those of household size. This differs sharply from the Ridker and Watson projection, in which newspapers are assumed to be largely replaced by electronic media. The cross-section data show no evidence of such replacement, and alternative media such as television have not yet eliminated the market for newspapers and magazines. For sectors 2 (poultry and eggs), 23 (meat products), 24 (dairy products), 28 (sugar), and 31 (soft drinks), the "income" variables (food at home and food away from home, and their respective log inverses) were held constant at the 1972 levels: this is equivalent to assuming saturation for each household type, allowing for only demographic effects after 1972. Sectors 2 and 23 are, however, allowed to change further by the subsequent imposition of a time trend.

For these two and for a number of other sectors, exogenous projections show a change over time in the response of expenditure to income growth. We introduced such trends by multiplying each of the demand equations in a sector (one for each household size class) by a term of the form $\exp(f(t))$, where $t=0$ in 1972 and $f(t)$ is chosen to accelerate or dampen the growth that would otherwise occur with income. This procedure was applied to the following sectors:

Sector No.	form of $f(t)$
2 (poultry and eggs)	$-0.007t + 0.0004\,(t)^{1.5}$
23 (meat products)	$0.0275(t)^{0.5}$
29 (confectionary products)	$-0.0066t$
32 (fats and oils)	$-0.0028t$
61 (miscellaneous chemicals)	$0.01t$
74 (miscellaneous plastics)	$0.016t$
125 (radio and television)	$0.006t$
146 (watches and clocks)	$-0.0078t$
147 (jewelry and silverware)	$-0.0175t$
148 (toys, sporting goods)	$0.0064t$
154 (water transport)	$-0.0159t$
158 (telephone and telegraph)	$0.0078t$
168 (real estate)	$0.0013t$

For food sectors (2, 23, 24, 28, 29, 31, and 32) these adjustments reflect the view that per capita saturation has already been reached, or will be reached by the end of the projection period, through a slowing or even a reversal of growth. For non-food sectors, the adjustments override the income effects in order to match the overall growth in scenario E to that derived from Ridker and Watson's equations, which are regarded as more reasonable, or to reflect saturation. Scenarios D and F are still allowed to differ from E, with demographic effects contributing much more to the difference than income effects. (No comparison was made to Ridker and Watson's D and F projections, since those take no account of composition effects.)

These multiplicative adjustments were calculated for the projections assuming continued income growth, so the functions $f(t)$ are defined by time rather than by income level or income changes. If the same adjustments were applied to the constant-productivity scenarios, they would, for many sectors, lead to <u>decline</u> in spending through time: this would

happen even though no reason had been identified why expenditures should
change. For this reason, the adjustments were <u>not</u> applied to the projections which assume constant income, it being assumed that the income coefficients estimated are correct so long as income changes relatively little,
but become erroneous for large changes. The only way to treat all projections alike would be to use more complicated demand functions, in which
the income coefficients would themselves be functions of income, and it
would not be necessary to introduce time as an instrumental variable to
bring about the changes. The short-cut we adopted, retaining the simple
form of the expenditure functions, does not vitiate comparisons between
growth and no-growth scenarios, because no independent assumptions about
time trends are introduced; "time" is only a proxy for income-related
changes. The difficulty with this approach arises from the fact that the
income projections in the "growth" scenarios follow the recession that
began in 1973, so that projected income actually declines during the middle
1970s before resuming growth. This factor together with the time-trend
adjustments means that for some sectors, paradoxically, expenditure is
projected to be lower when productivity increases than when it does not.
For most of the sectors affected, this problem occurs only out to 1985;
for sector 2 (poultry and eggs), however, the discrepancy persists to the
year 2025, because of the difference in how the adjustment (time-trend)
is applied.

When presenting the projection results, we show the expenditures
actually calculated. The use of these values in calculations of the sources and elasticities of expenditure growth would lead, for some sectors,
to negative effects. To avoid this result, we base the calculations on
the growth-scenario projections whenever the static-income projections
show higher expenditures. For most sectors, this adjustment is necessary
only for the 1975-1985 period; thereafter, the "growth" scenarios give
the higher expenditures and the projected values yield non-negative income
effects.

Launching the Projections

The regression equations, adjusted if necessary as described above,
may represent satisfactorily the <u>growth</u> in expenditure in a sector, and

yet not project the correct _level_ of spending in the base year. This can happen because of systematic under- or over-reporting by households (non-sampling errors in the surveys), compared to national accounts estimates of spending. Since the object is, in part to compare these new projections to those derived from aggregate data, it is essential that the projections pass through the same totals sector by sector in the base year. There-after, total PCE is equal in the two sets of projections, but the composition of expenditures can change.

This initial adjustment, required for "launching" the projections, is applied in the simplest possible way, by multiplying the 1972-73 survey total in each sector by a constant which will make it equal to the national accounts total, and then using that constant in all subsequent years. The assumption behind this once-for-all correction is that the degree of over- or under-reporting by households is independent of size, age, composition, and income, so that the adjustment does not need to be refined as income rises and the structure of the population changes. (Since national accounts estimates are not built up from estimates for groups of households, this assumption obviously cannot be checked.)

The adjustment factors are grouped in table 6-1, for 72 sectors which account for 42 percent of total PCE in the base year. (Other sectors either were not estimated from the survey data or required a different kind of adjustment.) Fifty-five percent of this amount is in sectors for which the factor lies between three-fourths and four-thirds. This range includes all the food sectors except 26 (grain products), 28 (sugar), and 29 (confectionary products). Sectors with extreme under- or over-estimation (factors below one-third or over three) take up 5 percent of PCE (12 percent of the total in the 72 sectors). These include some sectors in which few households buy anything and high accuracy is not to be expected, such as fertilizers, metal stampings, welding apparatus, x-ray equipment, measuring devices, and business services, and one sector--alcoholic beverages--notorious for systematic under-reporting by households.

Adjustments for Durable Goods

Cross-section household budget data may provide an acceptable explanation and a basis for projection of relatively frequent, regular, short-

Table 6-1. Adjustment Factors: Ratios of Sector Expenditure in National
Accounts to Expenditure Estimated from the CES, 1972-73

Range of Factor	Sectors	No. of sectors	Share of PCE from CES*	Total
Under 0.33	59, 87, 166	3	2.36	4.97
0.33 - 0.50	35, 158, 169, 170	4	4.64	9.77
0.50 - 0.75	26, 61, 81, 123, 126, 148	6	1.98	4.17
0.75 - 0.90	27, 25, 27, 32, 40, 72, 130, 145, 162, 168	11	9.33	19.64
0.90 - 1.00	152	1	0.41	0.86
1.00 - 1.11	24, 33, 56, 67, 103, 124, 153	7	3.58	7.54
1.11 - 1.33	23, 31, 38, 39, 49, 54, 74, 76, 154, 155, 157, 174, 177	13	9.85	20.74
1.33 - 2.00	28, 29, 37, 52, 58, 73, 80, 116, 125, 128, 139, 146, 147	13	2.63	5.54
2.00 - 3.00	34, 43, 53, 57, 77, 165	6	4.57	9.62
Over 3.00	30, 82, 97, 122, 131, 142, 151, 171	8	2.52	5.31
Total		72**	41.87	88.15

*Sectors estimated from the CES (including sectors 45 and 133 but not 163
and 164) account for 47.50 percent of total PCE.

**Excluding sectors 45 (furniture), 133 (automobiles), 140 (trailer coaches),
163 and 164 (trade margins) and 176 (private education).

term expenditures. They are much less reliable, however, for understanding or predicting purchases of durable goods. This is partly just because
such goods are purchased infrequently and expenditure may be concentrated
at particular stages of the life cycle, but there are more serious difficulties, two of which deserve mention. One is that there may be an income
threshold below which such goods are not bought, so that the Engel curve
should, even more than for other goods, have a logistic shape. The other
is that purchases in any one period depend not only on income and other
household characteristics during that period, but on the stock of durables
acquired in previous periods and the rate at which durables depreciate
with age or use. Stock information, or data on past purchases, could in

principle be included in cross-section surveys but usually is not, and might not be reliable as to the age or condition of durables in any case. In time-series data, stocks are not observed, but their effect can be modelled by a dynamic function, using as arguments income in the current period and expenditure on the goods in one or more previous periods. This is the approach taken by Almon and his associates in the original INFORUM model (1974), and also by Houthakker and Taylor (1971). The latter distinguish four categories of household durables, for which very good fits are obtained to observed annual expenditures; new cars and net purchases of used cars form a separate sector, for which the fit is less good but still quite satisfactory. The INFORUM model is concerned only with new car purchases, since used car transactions make no new demands on materials, and it distinguishes two furniture sectors (one for household and one for other use) and one for appliances. Cutlery, glassware, china, etc. are distributed over several sectors according to the material of which the items are made. We wanted to estimate two of these sectors in particular from the survey data, so as to capture demographic effects: These are 45 (furniture, which initially takes up 0.74 percent of PCE) and 133 (new automobiles, initially 4.88 percent of PCE). The other durable sectors presented no significant problem.

For both these sectors, equations can be estimated from the survey data with reasonable coefficients and acceptable overall fit. The values of the income coefficient (β_1) and the R^2 statistic are as follows, for the six household size classes:

Size	Sector 45 (furniture)		Sector 133 (new cars)	
	β_1	R^2	β_1	R^2
1	0.0193	0.1474	0.1119	0.1622
2	0.0136	0.1162	0.1095	0.1522
3	0.0124	0.0993	0.1237	0.1677
4	0.0212	0.1359	0.1775	0.2338
5	0.0122	0.1072	0.0522	0.1057
6 or more	0.0008	0.0631	0.0540	0.1478

Marginal spending propensities are (except in large households) between
one and two percent for furniture; for cars, they rise with household size
from 11 percent to 18 percent and then fall back to 5 percent. Demographic effects are clearly important, and this is reinforced by finding significant age and composition coefficients in several size classes.

The difficulty is that if these equations are used for projection,
they lead to unreasonably high total expenditures in the sectors before
the end of the projection period. The INFORUM projection for sector 133
in scenario E reaches 59.7 billion dollars in the year 2000, and 97.3 billion in 2025; our regression equations yield estimates of 85.8 billion and
171.4 billion respectively in the two years. Incomes are low enough now
to lead to a high marginal expenditure on automobiles, but if that rate is
sustained as income grows, the total number of new cars will become unreasonably large relative to the population. This imbalance could be broken
by a rise in price of cars sufficient to keep expenditure high while numbers of units decline; or by a sharp decline in the lifetime of a vehicle,
leading to much more frequent purchases. The second effect is unlikely,
and will be even less likely if relative car prices rise; and while the
first effect cannot be ruled out, it seems unlikely to be large enough.
Eventual saturation in expenditure as well as in physical units seems almost certain, and the intuitively most appealing level for saturation is
that rate of new car purchases which will maintain a stock of one car per
licensed driver or, to simplify somewhat, one car per person of driving
age. Given the size and structure of the population, the saturation level
of new car expenditures would then depend only on the average price and
the average lifetime of a car; the second is not expected to change dramatically from current levels. Real unit cost may be expected to rise
another 50 percent over the next half-century, at about the same rate as
in the period 1920-1970. This estimate of growth may be too rapid, however, because cars are now being made smaller and lighter, partly offsetting the trend to increased cost associated with other design changes.
Over the projection period, mass transit does not seem likely to reduce
significantly the desired stock of vehicles, although it may affect their
rate of use and thus expenditures on fuel and maintenance (Shapanka, 1978).

In an attempt to provide a better basis for projecting car purchases, we tried two procedures: modifying the survey-based equation, and finding a different demographically-sensitive projection model in the published literature. Two principal modifications to our equation were tried, both affecting the income coefficient. In one, the coefficient was reduced so as to minimize the sum of squares of differences between the projection using the CES equation and the INFORUM projection, differences being measured at five-year intervals over 1975-2025. In the other modification, a time trend was incorporated to pull down the income coefficient gradually, again with the objective of approximating the INFORUM results. Both these procedures were applied to scenario E; scenario D and F could then be projected using the same (modified) coefficients, so that demographic effects would be preserved. Neither procedure is very satisfactory, however. The first leads to a once-for-all reduction in the income effect, so that growth in demand is initially too slow but may be too fast toward the end of the period, while the time-trend adjustment is quite arbitrary and is not based on any explicit model of how saturation is approached. Both approaches are similar in principle to the introduction to time trends in other sectors, discussed earlier in this chapter, but with the disadvantages that there is no projection exogenous to the INFORUM equation to which to compare them.

The search for an alternative model for projection was no more successful; while many estimates of future automobile purchases have been made, none takes account of all the demographic factors we wish to consider for long-range projection. The most successful model appears to be that of Smith (1976), who uses a time-series of cross-section surveys to relate automobile purchases to ownership, via rates of scrapping and resale, and relates the probability of ownership to income level using a logistic function with 100 percent saturation for one car per household and saturation at slightly lower levels for second car ownership. His model separates new and used cars, which he finds to be very distant substitutes for most households, a result that is convenient for our purposes. His projections do not, however, depend on age, size, or composition of households, and it is not easy to predict which kinds of households are most likely to buy new cars.

A summary of other automobile forecasting models has been prepared by International Research and Technology Corporation (IR&T, 1976). IR&T has produced a sophisticated model of its own, using data from 12 countries to fit a logistic function between ownership and PCE per capita, but since our interest is only in the United States, this does not advance over Smith's more detailed work. The model projects market shares for small, medium, and large cars, so as to take account of price and quality changes, but, unfortunately, it includes no demographic variables and its projections stop at 1995. Nine other models are also available, but none uses the desired independent variables and all are subject to criticism as to their structural assumptions. It emerges from this review that the original INFORUM model of Almon et. al. is one of the most satisfactory for projection; it can also be refined to include price and depreciation effects.

Given the difficulties described, we decided to retain, for the basic projection of expenditure on new cars in scenario E, the INFORUM (Ridker-Watson) equation. The ratio of this projection to the expenditure obtained from the equation based on the survey defines a multiplicative factor for each year, which can then be applied to our CES projections for scenarios D and F. This procedure retains the demographic effects in the survey equation, but reduces their weight by a factor that depends on the year and the expenditure projected in scenario E. That is, absolute demographic differences are reduced, but relative effects are unchanged. The adjustment factor in the base year (whose value is 1.100) has the same meaning as for other sectors, correcting for survey misreporting. In later years, however, the adjustment also "corrects" for a "wrong" projection, and its value declines to 0.699 in 2000 and to 0.603 in 2025.

Furniture is the other durable good sector for which our unadjusted survey-based equations lead to implausibly high expenditures. In dollar terms the problem is less severe, but as a fraction of the time-series estimates it is comparable to that encountered for automobiles. The same procedure was therefore applied to sector 45 (furniture), using a multiplicative factor derived from scenario E to adjust projections for D and F in the same year. No adjustments were made to other durable goods sectors such as 103 (farm machinery), 116 (service industry machinery), 123 (household appliances), or 125 (radio and television receivers).

Sectoral Projections: Results

Projections were made using the income and household composition projections described in chapter 2, with the expenditure equations presented
in chapters 2 through 5, subject to the adjustments just described. The
results are presented in tables 6-2 through 6-7, which show for each sector in each scenario the expenditure (in millions of 1971 dollars) and
the share of total PCE. (These expenditures can be related to total consumption plus construction via table 1-2.) Tables 6-2 through 6-4 report
the results based on productivity growth for population scenarios D, E and
F respectively, and tables 6-5 through 6-7 show the results of the projections assuming no growth in productivity. The tables are arranged to
facilitate comparisons through time rather than among scenarios because
such comparisons offer more scope for analysis.

There are four sectors for which Ridker and Watson's projections lead
to zero expenditure by households before 2025. These are coal (sector 14),
ammunition and other ordnance (sectors 21 and 22) and plumbing and heating
equipment (sector 94). This does not mean the sectors would go out of
business, only that they would cease to sell anything directly to households. The assumption may be highly questionable except in the case of
coal for home heating, but the CES data do not permit us to make alternative projections. Ridker and Watson did not project expenditures in the
case of no productivity growth, so for our projections on this assumption,
we have usually retained their equations but held income at the base-year
level. For these four sectors, that approach leads to expenditure rising
as population increases; in tables 6-5 through 6-7, therefore, these four
projections are replaced by the corresponding estimates for the case of
productivity growth, drawn from tables 6-2 through 6-4.

A similar difficulty arises for several other sectors in which, because of adjustments or other factors which override the income effects,
expenditure is projected to be lower when income grows than when it does
not. The most striking examples are expenditures for petroleum deriva-
tives (69 and 70) and natural gas (161), where supply constraints are projected to push prices up and reduce consumption much below what it would
be if prices were constant. The same problem occurs for different reasons

Table 6-2
Expenditure and Percentage Share of PCE,
by Sector and Year, with Productivity Growth

Scenario D

Sector	1975 MLN 1971$	%PCE	1985 MLN 1971$	%PCE	2000 MLN 1971$	%PCE	2025 MLN 1971$	%PCE
1	75.000	0.01	50.000	0.01	50.000	0.00	50.000	0.00
2	1448.425	0.22	1611.893	0.17	1786.398	0.11	2165.139	0.07
3	264.410	0.04	358.126	0.04	573.177	0.04	1070.272	0.03
7	4288.961	0.64	5379.273	0.57	7420.102	0.48	11754.730	0.37
8	431.094	0.06	489.896	0.05	662.724	0.04	694.582	0.02
10	30.821	0.00	43.943	0.00	70.058	0.00	130.934	0.00
14	126.096	0.02	0.000	0.00	0.000	0.00	0.000	0.00
16	21.894	0.00	31.935	0.00	51.674	0.00	91.957	0.00
21	164.220	0.02	21.275	0.00	0.000	0.00	0.000	0.00
22	143.119	0.02	47.367	0.00	0.000	0.00	0.000	0.00
23	23117.766	3.44	27741.434	2.92	34301.383	2.20	47075.629	1.49
24	10955.422	1.63	12813.648	1.35	15021.008	0.96	19402.500	0.61
25	10538.230	1.57	13306.637	1.40	18241.496	1.17	28223.277	0.89
26	3530.187	0.52	4380.520	0.46	5863.863	0.38	8906.980	0.28
27	7565.066	1.12	9523.969	1.00	13168.184	0.85	20617.008	0.65
28	1110.681	0.17	1283.168	0.14	1488.951	0.10	1959.363	0.06
29	2904.061	0.43	3416.426	0.36	4284.551	0.28	5717.375	0.18
30	10808.480	1.61	14203.441	1.50	20534.570	1.32	33266.383	1.05
31	4855.738	0.72	5735.367	0.60	6796.316	0.44	8560.316	0.27
32	1185.216	0.18	1417.569	0.15	1758.271	0.11	2437.266	0.08
33	5574.348	0.83	7027.164	0.74	9546.504	0.61	14827.484	0.47
34	7445.348	1.11	9005.301	0.95	11210.523	0.72	14862.344	0.47
35	932.543	0.14	1156.413	0.12	1382.042	0.09	1889.876	0.06
36	1437.931	0.21	2298.683	0.24	3920.833	0.25	6304.273	0.20
37	149.527	0.02	184.955	0.02	247.205	0.02	367.992	0.01
38	3143.662	0.47	4204.105	0.44	6419.664	0.41	10955.918	0.35
39	15458.168	2.30	22455.387	2.36	40804.113	2.62	82563.312	2.61
40	2158.629	0.32	3316.030	0.35	6497.668	0.42	14216.086	0.45
41	188.404	0.03	231.167	0.02	333.401	0.02	535.509	0.02
43	283.627	0.04	421.535	0.04	806.627	0.05	1607.495	0.05
45	4689.379	0.70	6941.344	0.73	10662.035	0.68	20390.102	0.64
46	230.452	0.03	306.900	0.03	530.369	0.03	916.793	0.03
48	49.932	0.01	61.363	0.01	83.970	0.01	130.191	0.00
49	2269.157	0.34	2808.745	0.30	3645.211	0.23	5197.426	0.16
51	124.918	0.02	176.959	0.02	288.815	0.02	542.173	0.02
52	2165.660	0.32	2624.256	0.28	3111.408	0.20	4058.723	0.13
53	959.996	0.14	1440.605	0.15	2593.189	0.17	5146.836	0.16
54	1602.907	0.24	2085.132	0.22	3612.667	0.23	6844.055	0.22
55	244.425	0.04	637.875	0.07	1224.735	0.08	1441.385	0.05
56	100.444	0.01	121.661	0.01	153.867	0.01	219.516	0.01
57	207.519	0.03	277.991	0.03	381.665	0.02	585.937	0.02

Table 6-2 cont'd.

Sector	1975 MLN 1971$	%PCE	1985 MLN 1971$	%PCE	2000 MLN 1971$	%PCE	2025 MLN 1971$	%PCE
58	690.990	0.10	902.862	0.10	1242.313	0.08	1896.973	0.06
59	50.196	0.01	70.347	0.01	112.775	0.01	206.958	0.01
60	9.296	0.00	14.829	0.00	27.633	0.00	55.764	0.00
61	498.338	0.07	669.842	0.07	987.948	0.06	1801.076	0.06
62	22.655	0.00	35.072	0.00	62.937	0.00	123.533	0.00
66	8010.000	1.19	13316.000	1.40	25865.000	1.66	62942.000	1.99
67	6495.684	0.97	8053.422	0.85	10524.867	0.68	15284.426	0.48
68	34.753	0.01	47.528	0.01	72.473	0.00	126.715	0.00
69	10671.398	1.59	10567.125	1.11	10638.449	0.68	12766.566	0.40
70	2949.413	0.44	2585.359	0.27	2502.581	0.16	1036.873	0.03
72	2451.433	0.36	3216.195	0.34	4755.746	0.31	7566.195	0.24
73	791.204	0.12	979.523	0.10	1230.064	0.08	1714.073	0.05
74	490.182	0.07	708.055	0.07	1193.322	0.08	2586.974	0.08
76	3550.134	0.53	4804.148	0.51	7499.750	0.48	13214.082	0.42
77	991.673	0.15	1474.324	0.16	2601.067	0.17	5529.738	0.17
78	403.604	0.06	594.292	0.06	952.546	0.06	1654.892	0.05
80	137.135	0.02	187.342	0.02	237.863	0.02	393.864	0.01
81	4.019	0.00	5.105	0.00	6.906	0.00	9.647	0.00
82	157.971	0.02	183.838	0.02	225.063	0.01	318.989	0.01
83	9.710	0.00	12.214	0.00	19.215	0.00	25.729	0.00
87	18.659	0.00	23.667	0.00	30.870	0.00	44.897	0.00
90	2.558	0.00	3.685	0.00	5.301	0.00	5.731	0.00
93	0.542	0.00	4.614	0.00	10.372	0.00	8.716	0.00
94	21.571	0.00	0.000	0.00	0.000	0.00	0.000	0.00
95	39.321	0.01	62.987	0.01	122.653	0.01	245.904	0.01
96	28.855	0.00	41.710	0.00	76.563	0.00	135.516	0.00
97	476.277	0.07	588.187	0.06	712.185	0.05	997.212	0.03
98	775.367	0.12	1120.765	0.12	1843.277	0.12	3300.154	0.10
99	32.657	0.00	33.840	0.00	33.975	0.00	16.526	0.00
101	179.370	0.03	318.606	0.03	615.926	0.04	1251.054	0.04
102	218.187	0.03	260.332	0.03	469.524	0.03	892.748	0.03
103	579.541	0.09	792.108	0.08	1232.752	0.08	2128.443	0.07
106	35.736	0.01	56.326	0.01	98.062	0.01	178.600	0.01
108	108.911	0.02	182.552	0.02	320.578	0.02	592.042	0.02
109	28.839	0.00	36.650	0.00	62.538	0.00	114.901	0.00
115	170.132	0.03	287.705	0.03	562.937	0.04	1144.470	0.04
116	893.957	0.13	1193.752	0.13	1659.684	0.11	2500.373	0.08
117	5.475	0.00	8.924	0.00	17.722	0.00	36.974	0.00
119	12.332	0.00	18.614	0.00	37.761	0.00	73.125	0.00
120	20.175	0.00	34.246	0.00	70.527	0.00	137.383	0.00
122	1.047	0.00	1.344	0.00	1.942	0.00	3.094	0.00
123	5469.496	0.81	7449.020	0.78	11307.332	0.73	18856.762	0.60
124	756.775	0.11	1104.905	0.12	1951.108	0.13	3989.407	0.13
125	5117.492	0.76	7521.297	0.79	12889.570	0.83	25524.465	0.81
126	475.732	0.07	663.934	0.07	1060.387	0.07	1899.537	0.06
127	146.618	0.02	278.691	0.03	556.844	0.04	1139.004	0.04

Table 6-2 cont'd.

Sector	1975 MLN 1971$	%PCE	1985 MLN 1971$	%PCE	2000 MLN 1971$	%PCE	2025 MLN 1971$	%PCE
128	250.930	0.04	352.824	0.04	626.257	0.04	1156.505	0.04
129	471.792	0.07	643.969	0.07	1717.507	0.11	42061.137	1.33
130	165.423	0.02	211.651	0.02	295.088	0.02	438.327	0.01
131	90.880	0.01	116.968	0.01	167.343	0.01	263.089	0.01
133	27666.473	4.11	39466.160	4.16	59030.703	3.79	113831.625	3.59
134	278.545	0.04	479.950	0.05	985.147	0.06	2009.503	0.06
137	426.499	0.06	864.494	0.09	2059.309	0.13	4777.215	0.15
139	942.384	0.14	1314.966	0.14	2548.185	0.16	4913.219	0.16
140	1578.788	0.23	2681.555	0.28	3940.712	0.25	7478.562	0.24
142	25.068	0.00	32.048	0.00	49.763	0.00	73.813	0.00
143	464.800	0.07	768.100	0.08	1475.200	0.09	3568.101	0.11
144	232.400	0.03	384.050	0.04	737.600	0.05	1784.050	0.06
145	940.703	0.14	1361.255	0.14	2362.501	0.15	4447.617	0.14
146	510.256	0.08	761.729	0.08	1365.690	0.09	2488.095	0.08
147	1644.609	0.24	2551.146	0.27	5402.266	0.35	9545.395	0.30
148	3390.118	0.50	5310.340	0.56	9810.750	0.63	22265.977	0.70
149	285.168	0.04	410.921	0.04	690.046	0.04	1208.761	0.04
150	782.175	0.12	964.704	0.10	1482.870	0.10	2401.734	0.08
151	367.514	0.05	538.699	0.06	695.318	0.04	1330.378	0.04
152	2774.836	0.41	3570.524	0.38	5428.371	0.35	9804.934	0.31
153	670.331	0.10	1035.119	0.11	1759.796	0.11	3520.677	0.11
154	164.366	0.02	284.943	0.03	586.622	0.04	1050.108	0.03
155	3113.828	0.46	5526.051	0.58	12809.723	0.82	31509.102	0.99
157	15.346	0.00	23.365	0.00	30.336	0.00	43.614	0.00
158	11500.293	1.71	16751.609	1.76	27937.375	1.79	56823.242	1.79
160	14630.000	2.18	24667.082	2.60	36539.461	2.35	59511.746	1.88
161	5627.469	0.84	4813.324	0.51	3604.821	0.23	1069.690	0.03
162	2755.259	0.41	3610.729	0.38	5343.691	0.34	8920.152	0.28
163	39247.523	5.84	54964.879	5.79	91208.812	5.86	202118.500	6.38
164	117243.375	17.43	164195.563	17.29	272466.563	17.49	603785.375	19.05
165	19408.660	2.89	28033.465	2.95	53713.641	3.45	114137.000	3.60
166	15343.883	2.28	20937.066	2.20	33632.598	2.16	60292.676	1.90
167	72599.625	10.80	107473.562	11.32	181641.000	11.66	353671.063	11.16
168	32258.777	4.80	47763.586	5.03	73241.437	4.70	139236.250	4.39
169	3410.353	0.51	5592.168	0.59	13279.559	0.85	32696.070	1.03
170	13991.441	2.08	20743.582	2.18	37628.609	2.42	76942.812	2.43
171	4969.445	0.74	7468.539	0.79	13903.152	0.89	29449.180	0.93
172	250.357	0.04	345.926	0.04	640.162	0.04	1303.644	0.04
173	11861.695	1.76	15123.180	1.59	22715.398	1.46	39323.324	1.24
174	6196.391	0.92	10247.098	1.08	22455.578	1.44	52517.664	1.66
175	49037.797	7.29	81888.812	8.62	157073.188	10.09	379562.813	11.98
176	9283.000	1.38	13253.000	1.40	25005.000	1.61	37928.000	1.20
177	1784.877	0.27	2239.314	0.24	3027.670	0.19	4713.250	0.15
178	9.817	0.00	12.920	0.00	18.369	0.00	29.015	0.00
180	709.540	0.11	1066.957	0.11	1736.099	0.11	3330.760	0.11
181	229.957	0.03	1283.223	0.14	5253.187	0.34	13893.488	0.44

Table 6-3
Expenditure and Percentage Share of PCE,
by Sector and Year, with Productivity Growth

Scenario E

Sector	1975 MLN 1971$	%PCE	1985 MLN 1971$	%PCE	2000 MLN 1971$	%PCE	2025 MLN 1971$	%PCE
1	75.000	0.01	50.000	0.01	50.000	0.00	50.000	0.00
2	1445.088	0.22	1574.115	0.17	1693.762	0.11	1863.026	0.07
3	264.410	0.04	350.113	0.04	540.059	0.04	904.873	0.03
7	4269.262	0.64	5229.187	0.56	7195.062	0.47	10141.654	0.37
8	431.094	0.06	522.906	0.06	726.799	0.05	784.009	0.03
10	30.821	0.00	42.801	0.00	65.612	0.00	109.889	0.00
14	125.909	0.02	0.000	0.00	0.000	0.00	0.000	0.00
16	21.894	0.00	31.834	0.00	50.086	0.00	80.363	0.00
21	164.220	0.02	46.405	0.00	0.000	0.00	0.000	0.00
22	143.119	0.02	64.194	0.01	0.000	0.00	0.000	0.00
23	23026.367	3.44	26953.414	2.88	32338.160	2.11	40476.844	1.47
24	10932.098	1.63	12488.160	1.34	14150.035	0.92	16429.672	0.60
25	10495.883	1.57	12927.145	1.38	17537.887	1.14	23903.508	0.87
26	3518.117	0.53	4239.691	0.45	5570.648	0.36	7486.859	0.27
27	7532.035	1.13	9216.422	0.99	12614.793	0.82	17355.988	0.63
28	1108.353	0.17	1246.691	0.13	1400.403	0.09	1671.706	0.06
29	2892.779	0.43	3304.286	0.35	4096.875	0.27	4795.711	0.17
30	10768.070	1.61	13909.375	1.49	20108.520	1.31	28583.926	1.04
31	4843.371	0.72	5636.988	0.60	6456.207	0.42	7257.473	0.26
32	1181.549	0.18	1380.552	0.15	1685.792	0.11	2081.941	0.08
33	5555.324	0.83	6849.551	0.73	9254.586	0.60	12757.145	0.46
34	7434.617	1.11	8934.965	0.96	10908.277	0.71	12940.371	0.47
35	934.732	0.14	1161.561	0.12	1398.986	0.09	1662.549	0.06
36	1437.931	0.21	2337.005	0.25	3893.779	0.25	5670.035	0.21
37	148.643	0.02	180.372	0.02	238.470	0.02	318.644	0.01
38	3125.749	0.47	4055.488	0.43	6205.398	0.40	9137.461	0.33
39	15325.121	2.29	21518.437	2.30	40432.102	2.64	69437.750	2.53
40	2146.476	0.32	3183.627	0.34	6473.590	0.42	11818.586	0.43
41	188.404	0.03	230.153	0.02	324.950	0.02	476.149	0.02
43	281.255	0.04	404.628	0.04	797.717	0.05	1364.034	0.05
45	4658.922	0.70	6691.883	0.72	10667.293	0.70	17151.039	0.62
46	230.452	0.03	316.193	0.03	539.139	0.04	853.241	0.03
48	49.932	0.01	60.270	0.01	80.042	0.01	112.421	0.00
49	2262.273	0.34	2748.676	0.29	3502.826	0.23	4399.223	0.16
51	124.918	0.02	174.815	0.02	276.510	0.02	467.250	0.02
52	2164.844	0.32	2615.617	0.28	3039.648	0.20	3586.812	0.13
53	953.730	0.14	1433.557	0.15	2710.472	0.18	4639.223	0.17
54	1578.429	0.24	1948.944	0.21	3382.772	0.22	5480.102	0.20
55	244.438	0.04	626.702	0.07	1161.120	0.08	1212.309	0.04
56	100.101	0.01	116.455	0.01	142.393	0.01	175.186	0.01
57	207.843	0.03	281.156	0.03	387.592	0.03	525.284	0.02

Table 6-3 cont'd.

Sector	1975 MLN 1971$	%PCE	1985 MLN 1971$	%PCE	2000 MLN 1971$	%PCE	2025 MLN 1971$	%PCE
58	689.647	0.10	898.265	0.10	1223.431	0.08	1652.751	0.06
59	50.043	0.01	69.911	0.01	115.315	0.01	182.320	0.01
60	9.296	0.00	14.999	0.00	27.187	0.00	49.328	0.00
61	496.932	0.07	657.175	0.07	960.179	0.06	1572.692	0.06
62	22.655	0.00	35.337	0.00	61.722	0.00	109.065	0.00
66	7979.000	1.19	14006.000	1.50	26635.000	1.74	60339.000	2.19
67	6478.980	0.97	7903.824	0.84	10179.707	0.66	13074.867	0.48
68	34.753	0.01	46.712	0.00	68.915	0.00	108.464	0.00
69	10676.969	1.60	10554.980	1.13	10221.914	0.67	10952.262	0.40
70	2887.528	0.43	2498.438	0.27	2311.680	0.15	838.802	0.03
72	2435.048	0.36	3122.918	0.33	4608.566	0.30	6438.637	0.23
73	789.694	0.12	901.573	0.10	1055.574	0.07	1217.486	0.04
74	487.607	0.07	684.476	0.07	1132.235	0.07	2158.433	0.08
76	3530.868	0.53	4639.113	0.50	7320.891	0.48	11151.402	0.41
77	991.389	0.15	1441.165	0.15	2690.738	0.18	4720.914	0.17
78	403.604	0.06	587.826	0.06	911.418	0.06	1423.188	0.05
80	138.367	0.02	187.675	0.02	251.194	0.02	347.946	0.01
81	3.990	0.00	4.930	0.00	6.666	0.00	7.995	0.00
82	156.926	0.02	171.417	0.02	197.040	0.01	243.734	0.01
83	9.710	0.00	13.177	0.00	20.936	0.00	27.110	0.00
87	18.683	0.00	23.595	0.00	30.855	0.00	39.302	0.00
90	2.558	0.00	3.868	0.00	5.576	0.00	5.923	0.00
93	0.542	0.00	4.594	0.00	9.965	0.00	7.477	0.00
94	21.571	0.00	0.000	0.00	0.000	0.00	0.000	0.00
95	39.321	0.01	64.777	0.01	122.755	0.01	221.082	0.01
96	28.855	0.00	43.845	0.00	79.151	0.01	128.003	0.00
97	476.858	0.07	565.895	0.06	632.561	0.04	743.943	0.03
98	775.367	0.12	1113.892	0.12	1776.611	0.12	2863.276	0.10
99	32.657	0.00	34.608	0.00	35.908	0.00	22.228	0.00
101	179.370	0.03	321.774	0.03	602.691	0.04	1098.226	0.04
102	218.187	0.03	275.977	0.03	494.915	0.03	861.002	0.03
103	577.369	0.09	770.175	0.08	1206.818	0.08	1810.372	0.07
106	35.736	0.01	57.921	0.01	99.027	0.01	163.721	0.01
108	108.911	0.02	183.368	0.02	312.540	0.02	519.411	0.02
109	28.839	0.00	38.012	0.00	64.213	0.00	107.678	0.00
115	170.132	0.03	294.594	0.03	560.539	0.04	1024.197	0.04
116	891.026	0.13	1168.556	0.12	1628.715	0.11	2122.757	0.08
117	5.475	0.00	9.100	0.00	17.544	0.00	32.827	0.00
119	12.332	0.00	20.028	0.00	39.794	0.00	69.840	0.00
120	20.175	0.00	36.534	0.00	73.370	0.00	129.140	0.00
122	1.044	0.00	1.234	0.00	1.693	0.00	2.278	0.00
123	5447.395	0.81	7307.473	0.78	11164.523	0.73	16150.332	0.59
124	754.887	0.11	1070.493	0.11	1943.268	0.13	3362.963	0.12
125	5089.230	0.76	7332.117	0.78	12621.195	0.82	21656.437	0.79
126	473.326	0.07	647.001	0.07	1033.374	0.07	1574.639	0.06
127	146.618	0.02	289.178	0.03	563.363	0.04	1037.803	0.04

Table 6-3 cont'd.

Sector	1975 MLN 1971$	%PCE	1985 MLN 1971$	%PCE	2000 MLN 1971$	%PCE	2025 MLN 1971$	%PCE
128	248.210	0.04	338.626	0.04	611.433	0.04	963.512	0.04
129	471.792	0.07	651.039	0.07	1657.647	0.11	34905.637	1.27
130	164.364	0.02	206.189	0.02	285.517	0.02	376.473	0.01
131	90.667	0.01	108.399	0.01	147.731	0.01	196.001	0.01
133	27240.168	4.07	37690.758	4.03	59699.852	3.89	97266.375	3.54
134	278.545	0.04	526.831	0.06	1067.771	0.07	1978.574	0.07
137	426.499	0.06	879.779	0.09	2037.371	0.13	4253.984	0.15
139	924.825	0.14	1226.831	0.13	2372.437	0.15	4009.573	0.15
140	1605.942	0.24	2638.685	0.28	3513.116	0.23	5422.543	0.20
142	24.787	0.00	31.353	0.00	48.761	0.00	64.259	0.00
143	463.400	0.07	808.300	0.09	1509.900	0.10	3380.901	0.12
144	231.700	0.03	404.150	0.04	754.950	0.05	1690.450	0.06
145	933.698	0.14	1322.799	0.14	2331.771	0.15	3741.429	0.14
146	506.624	0.08	735.392	0.08	1392.608	0.09	2102.906	0.08
147	1623.795	0.24	2401.228	0.26	5469.168	0.36	8002.379	0.29
148	3374.296	0.50	5050.891	0.54	9315.004	0.61	17657.309	0.64
149	285.169	0.04	411.428	0.04	673.558	0.04	1066.800	0.04
150	782.175	0.12	984.852	0.11	1497.745	0.10	2231.711	0.08
151	371.389	0.06	539.209	0.06	711.884	0.05	1164.033	0.04
152	2755.223	0.41	3467.237	0.37	5327.559	0.35	8483.723	0.31
153	669.284	0.10	1028.776	0.11	1885.813	0.12	3205.491	0.12
154	164.555	0.02	283.586	0.03	645.922	0.04	917.876	0.03
155	3076.448	0.46	5271.359	0.56	13252.797	0.86	27135.273	0.99
157	15.295	0.00	23.833	0.00	32.683	0.00	40.434	0.00
158	11460.105	1.71	16512.348	1.77	27710.383	1.81	49256.215	1.79
160	14292.738	2.14	23840.687	2.55	33775.645	2.20	48421.062	1.76
161	5720.047	0.85	4651.496	0.50	3329.841	0.22	865.356	0.03
162	2744.553	0.41	3514.183	0.38	5193.078	0.34	7632.191	0.28
163	39042.656	5.83	54185.645	5.79	90237.562	5.89	175058.813	6.37
164	116631.375	17.42	161867.688	17.30	269565.125	17.59	522949.938	19.02
165	19199.398	2.87	26553.883	2.84	52448.570	3.42	96276.062	3.50
166	15241.430	2.28	20340.023	2.17	33249.102	2.17	51761.621	1.88
167	72599.625	10.85	105662.562	11.30	172417.688	11.25	301319.000	10.96
168	32288.809	4.82	48287.410	5.16	74851.750	4.88	125043.437	4.55
169	3343.612	0.50	5204.012	0.56	13488.367	0.88	27612.035	1.00
170	13895.602	2.08	20130.016	2.15	37981.656	2.48	66029.375	2.40
171	4935.457	0.74	7272.180	0.78	14102.520	0.92	25187.555	0.92
172	250.357	0.04	350.884	0.04	633.305	0.04	1161.390	0.04
173	11861.695	1.77	14970.125	1.60	21976.406	1.43	34436.863	1.25
174	6120.262	0.91	9660.477	1.03	22421.660	1.46	43589.297	1.59
175	48640.898	7.27	85529.500	9.14	159698.125	10.42	356410.625	12.96
176	9634.000	1.44	15385.000	1.64	26555.000	1.73	53458.000	1.94
177	1780.575	0.27	2218.241	0.24	3001.568	0.20	4215.441	0.15
178	9.817	0.00	12.779	0.00	17.684	0.00	25.302	0.00
180	709.540	0.11	1030.942	0.11	1605.336	0.10	2753.428	0.10
181	229.957	0.03	1472.684	0.16	5273.402	0.34	11456.559	0.42

Table 6-4
Expenditure and Percentage Share of PCE,
by Sector and Year, with Productivity Growth

Scenario F

Sector	1975 MLN 1971$	%PCE	1985 MLN 1971$	%PCE	2000 MLN 1971$	%PCE	2025 MLN 1971$	%PCE
1	75.000	0.01	50.000	0.01	50.000	0.00	50.000	0.00
2	1451.687	0.22	1552.206	0.16	1671.193	0.11	1707.074	0.07
3	264.410	0.04	347.263	0.04	521.304	0.03	808.680	0.03
7	4279.414	0.64	5225.094	0.55	7122.043	0.47	9368.109	0.37
8	431.094	0.06	576.705	0.06	798.728	0.05	875.187	0.03
10	30.821	0.00	42.327	0.00	62.985	0.00	97.387	0.00
14	125.816	0.02	0.000	0.00	0.000	0.00	0.000	0.00
16	21.894	0.00	32.321	0.00	49.654	0.00	74.202	0.00
21	164.220	0.02	75.444	0.01	0.000	0.00	0.000	0.00
22	143.119	0.02	83.362	0.01	3.517	0.00	0.000	0.00
23	23206.926	3.47	26817.723	2.82	31955.371	2.09	37019.574	1.45
24	10977.879	1.64	12241.855	1.29	13916.387	0.91	14937.379	0.59
25	10507.984	1.57	12848.367	1.35	17306.312	1.13	21880.301	0.86
26	3523.365	0.53	4171.184	0.44	5498.848	0.36	6806.055	0.27
27	7547.734	1.13	9135.328	0.96	12428.855	0.81	15839.445	0.62
28	1112.891	0.17	1215.091	0.13	1376.912	0.09	1536.939	0.06
29	2895.350	0.43	3259.866	0.34	4038.427	0.26	4380.781	0.17
30	10762.457	1.61	13961.531	1.47	19880.848	1.30	26241.754	1.03
31	4867.473	0.73	5621.082	0.59	6360.691	0.42	6504.203	0.25
32	1182.951	0.18	1364.028	0.14	1663.150	0.11	1919.036	0.08
33	5560.285	0.83	6800.352	0.72	9142.547	0.60	11786.945	0.46
34	7446.316	1.11	8980.477	0.95	10813.824	0.71	11772.332	0.46
35	925.523	0.14	1115.142	0.12	1366.886	0.09	1553.527	0.06
36	1437.931	0.21	2432.902	0.26	3947.208	0.26	5368.793	0.21
37	149.627	0.02	182.765	0.02	236.114	0.02	291.787	0.01
38	3122.958	0.47	4051.531	0.43	6121.246	0.40	8338.359	0.33
39	15288.727	2.29	21854.559	2.30	40008.109	2.62	63996.418	2.51
40	2115.889	0.32	3185.204	0.34	6431.691	0.42	10907.367	0.43
41	188.404	0.03	233.353	0.02	323.759	0.02	446.962	0.02
43	277.761	0.04	419.125	0.04	799.370	0.05	1254.611	0.05
45	4597.359	0.69	6758.461	0.71	10593.383	0.69	15762.203	0.62
46	230.453	0.03	330.375	0.03	556.059	0.04	836.141	0.03
48	49.932	0.01	60.181	0.01	78.139	0.01	102.584	0.00
49	2265.886	0.34	2737.919	0.29	3458.864	0.23	3991.350	0.16
51	124.918	0.02	175.536	0.02	270.983	0.02	425.736	0.02
52	2166.308	0.32	2607.064	0.27	3004.629	0.20	3295.179	0.13
53	947.314	0.14	1498.406	0.16	2713.458	0.18	4329.160	0.17
54	1599.256	0.24	2049.766	0.22	3335.083	0.22	4902.520	0.19
55	244.454	0.04	620.517	0.07	1122.645	0.07	1066.272	0.04
56	100.385	0.02	111.588	0.01	138.083	0.01	156.466	0.01
57	206.030	0.03	283.899	0.03	385.621	0.03	490.225	0.02

Table 6-4 cont'd.

Sector	1975 MLN 1971$	%PCE	1985 MLN 1971$	%PCE	2000 MLN 1971$	%PCE	2025 MLN 1971$	%PCE
58	688.219	0.10	912.673	0.10	1215.553	0.08	1524.032	0.06
59	49.701	0.01	71.289	0.01	115.196	0.01	169.684	0.01
60	9.296	0.00	15.425	0.00	27.299	0.00	46.138	0.00
61	498.344	0.07	654.308	0.07	947.949	0.06	1446.059	0.06
62	22.655	0.00	36.224	0.00	61.810	0.00	101.791	0.00
66	8012.000	1.20	14821.000	1.56	28244.000	1.85	59632.000	2.34
67	6484.590	0.97	7872.289	0.83	10061.141	0.66	11928.332	0.47
68	34.753	0.01	46.678	0.00	67.123	0.00	98.104	0.00
69	10680.551	1.60	10542.062	1.11	9984.660	0.65	9726.152	0.38
70	2937.846	0.44	2447.559	0.26	2189.548	0.14	735.969	0.03
72	2439.543	0.36	3179.516	0.33	4557.887	0.30	5882.227	0.23
73	786.808	0.12	798.387	0.08	1007.776	0.07	1065.623	0.04
74	490.062	0.07	681.120	0.07	1115.836	0.07	1948.520	0.08
76	3523.662	0.53	4627.590	0.49	7228.332	0.47	10234.223	0.40
77	968.747	0.14	1370.828	0.14	2639.265	0.17	4476.145	0.18
78	403.604	0.06	595.166	0.06	894.058	0.06	1289.405	0.05
80	132.879	0.02	161.476	0.02	238.147	0.02	345.653	0.01
81	4.007	0.00	4.893	0.00	6.532	0.00	7.188	0.00
82	158.422	0.02	163.803	0.02	190.455	0.01	218.708	0.01
83	9.710	0.00	14.685	0.00	22.894	0.00	28.985	0.00
87	18.525	0.00	23.031	0.00	30.453	0.00	36.490	0.00
90	2.558	0.00	4.263	0.00	5.961	0.00	6.158	0.00
93	0.542	0.00	4.586	0.00	9.733	0.00	6.640	0.00
94	21.571	0.00	0.000	0.00	0.000	0.00	0.000	0.00
95	39.321	0.01	67.748	0.01	125.091	0.01	209.978	0.01
96	28.855	0.00	47.505	0.01	83.058	0.01	126.270	0.00
97	472.196	0.07	524.147	0.06	606.545	0.04	662.409	0.03
98	775.367	0.12	1131.501	0.12	1753.677	0.11	2619.820	0.10
99	32.657	0.00	36.387	0.00	38.365	0.00	27.205	0.00
101	179.370	0.03	332.348	0.03	602.931	0.04	1017.295	0.04
102	218.187	0.03	295.806	0.03	525.441	0.03	871.190	0.03
103	573.153	0.09	764.079	0.08	1200.354	0.08	1655.048	0.06
106	35.736	0.01	60.799	0.01	101.748	0.01	158.037	0.01
108	108.911	0.02	189.082	0.02	311.924	0.02	479.835	0.02
109	28.839	0.00	39.964	0.00	66.793	0.00	106.299	0.00
115	170.132	0.03	307.865	0.03	569.095	0.04	967.271	0.04
116	885.591	0.13	1147.588	0.12	1597.635	0.10	1957.097	0.08
117	5.475	0.00	9.424	0.00	17.708	0.00	30.824	0.00
119	12.332	0.00	21.989	0.00	42.321	0.00	69.821	0.00
120	20.175	0.00	39.960	0.00	77.298	0.01	127.199	0.00
122	1.034	0.00	1.128	0.00	1.647	0.00	2.004	0.00
123	5424.691	0.81	7373.301	0.78	11068.832	0.72	14778.281	0.58
124	745.997	0.11	1058.156	0.11	1930.282	0.13	3101.924	0.12
125	5076.000	0.76	7448.137	0.78	12483.121	0.82	19772.562	0.77
126	470.765	0.07	652.408	0.07	1018.465	0.07	1429.924	0.06
127	146.618	0.02	306.378	0.03	579.741	0.04	996.360	0.04

Table 6-4 cont'd.

Sector	1975 MLN 1971$	%PCE	1985 MLN 1971$	%PCE	2000 MLN 1971$	%PCE	2025 MLN 1971$	%PCE
128	247.700	0.04	353.686	0.04	608.383	0.04	869.444	0.03
129	471.792	0.07	669.552	0.07	1636.415	0.11	30566.461	1.20
130	164.889	0.02	209.103	0.02	282.345	0.02	343.808	0.01
131	89.741	0.01	99.645	0.01	143.807	0.01	172.860	0.01
133	27022.227	4.04	39627.371	4.17	59556.820	3.89	89761.750	3.52
134	278.545	0.04	581.839	0.06	1157.482	0.08	2035.957	0.08
137	426.499	0.06	913.482	0.10	2056.950	0.13	3998.947	0.16
139	940.009	0.14	1347.084	0.14	2372.708	0.16	3523.496	0.14
140	1598.843	0.24	2808.594	0.30	3835.946	0.25	6005.852	0.24
142	25.042	0.00	33.851	0.00	48.803	0.00	57.252	0.00
143	466.400	0.07	855.900	0.09	1594.800	0.10	3310.801	0.13
144	233.200	0.03	427.950	0.05	797.400	0.05	1655.400	0.06
145	930.794	0.14	1364.125	0.14	2319.497	0.15	3410.077	0.13
146	499.441	0.07	737.787	0.08	1375.410	0.09	1957.448	0.08
147	1587.803	0.24	2467.622	0.26	5496.109	0.36	7353.758	0.29
148	3347.632	0.50	4871.453	0.51	9109.020	0.60	15965.062	0.63
149	285.170	0.04	417.995	0.04	671.137	0.04	995.145	0.04
150	782.175	0.12	1027.049	0.11	1539.509	0.10	2178.718	0.09
151	359.800	0.05	480.016	0.05	685.431	0.04	1163.043	0.05
152	2780.034	0.42	3580.505	0.38	5268.578	0.34	7873.977	0.31
153	659.323	0.10	1012.304	0.11	1857.257	0.12	3085.413	0.12
154	153.186	0.02	272.186	0.03	651.448	0.04	871.536	0.03
155	3025.729	0.45	5374.570	0.57	13191.160	0.86	25511.367	1.00
157	15.186	0.00	23.872	0.00	31.652	0.00	38.258	0.00
158	11423.441	1.71	16693.570	1.76	27481.184	1.80	45416.992	1.78
160	14536.668	2.17	23373.852	2.46	32034.273	2.09	42861.359	1.68
161	5598.645	0.84	4556.770	0.48	3153.916	0.21	759.266	0.03
162	2738.881	0.41	3522.858	0.37	5161.016	0.34	7010.914	0.27
163	38970.184	5.83	55035.512	5.80	90505.375	5.92	162982.375	6.39
164	116414.875	17.40	164406.563	17.31	270365.250	17.67	486874.500	19.08
165	19213.266	2.87	27213.391	2.87	52291.387	3.42	87663.562	3.44
166	15232.582	2.28	20709.367	2.18	32989.055	2.16	47692.852	1.87
167	72599.625	10.85	105623.125	11.12	167656.500	10.96	271242.500	10.63
168	32108.086	4.80	48683.387	5.13	73769.687	4.82	116141.687	4.55
169	3353.668	0.50	5444.363	0.57	13332.094	0.87	25778.043	1.01
170	13823.516	2.07	20510.172	2.16	37708.508	2.46	61474.590	2.41
171	4903.469	0.73	7420.391	0.78	13995.410	0.91	23230.973	0.91
172	250.357	0.04	361.075	0.04	638.681	0.04	1094.647	0.04
173	11861.695	1.77	15059.789	1.59	21720.520	1.42	31948.227	1.25
174	6065.512	0.91	9820.629	1.03	22188.020	1.45	40321.773	1.58
175	48696.398	7.28	90201.125	9.50	167726.750	10.96	347384.750	13.61
176	9723.000	1.45	15210.000	1.60	24790.000	1.62	43623.000	1.71
177	1780.672	0.27	2248.070	0.24	2988.483	0.20	3938.919	0.15
178	9.817	0.00	12.863	0.00	17.422	0.00	23.312	0.00
180	709.540	0.11	1010.000	0.11	1521.899	0.10	2400.865	0.09
181	229.957	0.03	1806.848	0.19	5524.969	0.36	10346.008	0.41

Table 6-5
Expenditure and Percentage Share of PCE,
by Sector and Year, without Productivity Growth

Scenario D

Sector	1975 MLN 1971$	%PCE	1985 MLN 1971$	%PCE	2000 MLN 1971$	%PCE	2025 MLN 1971$	%PCE
1	110.970	0.02	125.715	0.02	147.378	0.01	189.396	0.01
2	1472.691	0.21	1692.056	0.20	1941.940	0.19	2313.737	0.17
3	256.788	0.04	290.908	0.04	341.037	0.03	438.268	0.03
7	4478.590	0.63	5130.801	0.62	6189.574	0.60	7729.559	0.57
8	628.860	0.09	712.420	0.09	835.181	0.08	1073.295	0.08
10	30.080	0.00	34.077	0.00	39.949	0.00	51.339	0.00
14	160.559	0.02	0.000	0.00	0.000	0.00	0.000	0.00
16	22.194	0.00	25.143	0.00	29.475	0.00	37.879	0.00
21	274.431	0.04	21.275	0.00	0.000	0.00	0.000	0.00
22	197.091	0.03	47.367	0.01	0.000	0.00	0.000	0.00
23	21955.297	3.07	24409.137	2.94	27958.801	2.70	33132.902	2.46
24	10932.098	1.53	12488.160	1.50	14150.035	1.37	16429.672	1.22
25	11014.051	1.54	12688.488	1.53	15182.652	1.47	18367.500	1.36
26	3644.708	0.51	4181.066	0.50	4947.695	0.48	5942.496	0.44
27	7892.094	1.10	9047.340	1.09	10829.844	1.05	12987.469	0.96
28	1108.353	0.15	1246.691	0.15	1400.403	0.14	1671.706	0.12
29	3096.690	0.43	3532.463	0.43	4218.570	0.41	5066.992	0.38
30	11413.437	1.60	13593.105	1.64	16745.641	1.62	20195.578	1.50
31	4843.371	0.68	5636.988	0.68	6456.207	0.62	7257.473	0.54
32	1233.299	0.17	1412.879	0.17	1651.219	0.16	2030.033	0.15
33	5798.469	0.81	6734.445	0.81	8097.520	0.78	9987.879	0.74
34	7507.699	1.05	8902.613	1.07	10612.621	1.02	12289.078	0.91
35	962.958	0.13	1149.092	0.14	1298.126	0.13	1462.815	0.11
36	1626.041	0.23	1842.099	0.22	2159.523	0.21	2775.213	0.21
37	153.903	0.02	177.931	0.02	214.961	0.02	264.221	0.02
38	3406.050	0.48	3924.515	0.47	4853.465	0.47	5893.156	0.44
39	17633.441	2.46	20372.148	2.45	26904.852	2.60	34116.457	2.53
40	2543.752	0.36	2986.675	0.36	3997.779	0.39	5180.617	0.38
41	197.097	0.03	223.286	0.03	261.762	0.03	336.391	0.02
43	335.217	0.05	380.016	0.05	510.491	0.05	685.734	0.05
45	5130.277	0.72	5579.945	0.67	7376.645	0.71	10192.973	0.76
46	257.773	0.04	292.024	0.04	342.344	0.03	439.948	0.03
48	48.616	0.01	55.075	0.01	64.566	0.01	82.974	0.01
49	2332.024	0.33	2718.007	0.33	3214.829	0.31	3752.680	0.28
51	118.692	0.02	134.463	0.02	157.633	0.02	.202.575	0.02
52	2164.844	0.30	2615.617	0.31	3039.648	0.29	3586.812	0.27
53	1088.970	0.15	1363.702	0.16	1869.805	0.18	2451.096	0.18
54	1792.594	0.25	1852.002	0.22	2306.035	0.22	2824.179	0.21
55	208.568	0.03	236.282	0.03	276.997	0.03	355.970	0.03
56	102.671	0.01	115.315	0.01	129.865	0.01	144.147	0.01
57	216.911	0.03	276.790	0.03	344.870	0.03	424.501	0.03

Table 6-5 cont'd.

Sector	1975 MLN 1971$	%PCE	1985 MLN 1971$	%PCE	2000 MLN 1971$	%PCE	2025 MLN 1971$	%PCE
58	720.639	0.10	884.214	0.11	1085.850	0.10	1337.771	0.10
59	54.742	0.01	67.544	0.01	87.837	0.01	112.772	0.01
60	8.569	0.00	9.708	0.00	11.381	0.00	14.625	0.00
61	491.897	0.07	572.014	0.07	673.883	0.07	808.881	0.06
62	27.207	0.00	30.822	0.00	36.133	0.00	46.434	0.00
66	8010.000	1.12	9136.000	1.10	10683.000	1.03	14466.000	1.07
67	6673.574	0.93	7814.223	0.94	9271.543	0.89	10924.359	0.81
68	33.328	0.00	37.757	0.00	44.263	0.00	56.882	0.00
69	11134.863	1.56	12614.402	1.52	14788.070	1.43	19004.211	1.41
70	3083.625	0.43	3493.359	0.42	4095.322	0.40	5262.914	0.39
72	2637.545	0.37	3030.025	0.36	3764.810	0.36	4607.684	0.34
73	828.093	0.12	888.104	0.11	961.232	0.09	1047.586	0.08
74	481.870	0.07	548.527	0.07	651.727	0.06	761.741	0.06
76	3866.049	0.54	4478.238	0.54	5590.590	0.54	6859.078	0.51
77	1161.211	0.16	1356.322	0.16	1706.709	0.16	2135.771	0.16
78	476.628	0.07	539.960	0.06	633.004	0.06	813.476	0.06
80	156.346	0.02	179.539	0.02	189.898	0.02	226.330	0.02
81	4.246	0.00	4.826	0.00	5.832	0.00	6.465	0.00
82	161.523	0.02	169.465	0.02	181.374	0.02	211.129	0.02
83	15.476	0.00	17.533	0.00	20.554	0.00	26.414	0.00
87	19.428	0.00	23.259	0.00	27.777	0.00	32.564	0.00
90	4.624	0.00	5.239	0.00	6.141	0.00	7.892	0.00
93	0.405	0.00	0.459	0.00	0.538	0.00	0.692	0.00
94	48.226	0.01	0.000	0.00	0.000	0.00	0.000	0.00
95	39.962	0.01	45.272	0.01	53.073	0.01	68.204	0.01
96	49.478	0.01	56.052	0.01	65.711	0.01	84.445	0.01
97	500.180	0.07	558.547	0.07	589.632	0.06	652.415	0.05
98	910.084	0.13	1031.011	0.12	1208.671	0.12	1553.268	0.12
99	44.565	0.01	50.486	0.01	59.186	0.01	76.060	0.01
101	199.323	0.03	225.808	0.03	264.718	0.03.	340.190	0.03
102	254.379	0.04	288.179	0.03	337.837	0.03	434.156	0.03
103	633.389	0.09	744.821	0.09	941.486	0.09	1163.622	0.09
106	36.343	0.01	41.172	0.00	48.267	0.00	62.028	0.00
108	133.763	0.02	151.536	0.02	177.649	0.02	228.297	0.02
109	37.054	0.01	41.978	0.01	49.211	0.00	63.242	0.00
115	178.226	0.02	201.908	0.02	236.700	0.02	304.184	0.02
116	966.474	0.14	1135.971	0.14	1371.979	0.13	1626.120	0.12
117	5.209	0.00	5.901	0.00	6.918	0.00	8.890	0.00
119	15.666	0.00	17.748	0.00	20.806	0.00	26.738	0.00
120	28.607	0.00	32.408	0.00	37.992	0.00	48.824	0.00
122	1.147	0.00	1.192	0.00	1.332	0.00	1.541	0.00
123	5916.145	0.83	7086.445	0.85	8917.437	0.86	10796.820	0.80
124	850.406	0.12	1021.149	0.12	1313.227	0.13	1647.164	0.12
125	5488.297	0.77	6550.336	0.79	8290.070	0.80	10114.715	0.75
126	526.074	0.07	622.115	0.07	768.924	0.07	917.098	0.07
127	171.298	0.02	194.059	0.02	227.498	0.02	292.359	0.02

Table 6-5 cont'd.

Sector	1975 MLN 1971$	%PCE	1985 MLN 1971$	%PCE	2000 MLN 1971$	%PCE	2025 MLN 1971$	%PCE
128	287.409	0.04	321.165	0.04	414.796	0.04	509.854	0.04
129	502.782	0.07	569.589	0.07	667.738	0.06	858.114	0.06
130	175.276	0.02	201.402	0.02	246.301	0.02	297.582	0.02
131	99.057	0.01	104.961	0.01	118.168	0.01	135.671	0.01
133	34017.594	4.75	36776.051	4.43	51691.141	4.99	75286.562	5.59
134	255.259	0.04	289.176	0.03	339.006	0.03	435.658	0.03
137	511.768	0.07	579.769	0.07	679.672	0.07	873.450	0.06
139	1070.803	0.15	1159.851	0.14	1602.016	0.15	2050.276	0.15
140	1679.675	0.23	2377.789	0.29	2553.461	0.25	3305.866	0.25
142	26.881	0.00	30.408	0.00	40.369	0.00	47.466	0.00
143	464.800	0.06	527.000	0.06	609.300	0.06	820.000	0.06
144	232.400	0.03	263.500	0.03	304.650	0.03	410.000	0.03
145	1057.230	0.15	1263.253	0.15	1667.300	0.16	2063.039	0.15
146	627.458	0.09	758.436	0.09	1029.810	0.10	1311.901	0.10
147	2309.050	0.32	2690.990	0.32	4125.078	0.40	5726.887	0.43
148	3667.247	0.51	4261.121	0.51	5227.176	0.50	6067.820	0.45
149	299.254	0.04	339.018	0.04	397.436	0.04	510.747	0.04
150	915.203	0.13	1036.809	0.12	1215.469	0.12	1562.004	0.12
151	401.453	0.06	523.505	0.06	574.523	0.06	845.851	0.06
152	2902.485	0.41	3386.035	0.41	4222.531	0.41	5402.016	0.40
153	765.041	0.11	975.995	0.12	1280.925	0.12	1640.518	0.12
154	248.718	0.03	321.074	0.04	477.456	0.05	647.645	0.05
155	4028.228	0.56	4775.723	0.57	6973.352	0.67	9976.887	0.74
157	16.957	0.00	23.028	0.00	26.982	0.00	30.694	0.00
158	12011.000	1.68	14525.770	1.75	18129.875	1.75	22409.656	1.66
160	12240.273	1.71	13866.687	1.67	16256.148	1.57	20890.844	1.55
161	6219.555	0.87	7045.977	0.85	8260.113	0.80	10615.105	0.79
162	2935.469	0.41	3425.134	0.41	4260.105	0.41	5337.113	0.40
163	42597.922	5.95	50292.168	6.05	65078.371	6.28	89014.187	6.61
164	127251.937	17.78	150236.875	18.08	194407.438	18.76	265910.625	19.75
165	22200.633	3.10	25037.863	3.01	33639.980	3.25	43391.996	3.22
166	16805.020	2.35	19587.766	2.36	25105.758	2.42	31537.582	2.34
167	72894.062	10.19	82579.812	9.94	96809.687	9.34	124410.562	9.24
168	32742.184	4.58	43398.164	5.22	52946.641	5.11	62506.324	4.64
169	4367.766	0.61	4674.344	0.56	6878.430	0.66	9832.070	0.73
170	15919.012	2.22	19101.297	2.30	25623.859	2.47	33461.852	2.49
171	5681.488	0.79	6884.758	0.83	9083.629	0.88	11350.414	0.84
172	240.139	0.03	272.047	0.03	318.925	0.03	409.852	0.03
173	12247.059	1.71	13874.375	1.67	16265.160	1.57	20902.430	1.55
174	7699.113	1.08	8860.797	1.07	12473.238	1.20	16605.660	1.23
175	49037.797	6.85	56186.500	6.76	64874.047	6.26	87233.000	6.48
176	9061.000	1.27	10349.000	1.25	11683.000	1.13	15051.000	1.12
177	1844.598	0.26	2186.670	0.26	2663.318	0.26	3388.673	0.25
178	10.051	0.00	11.386	0.00	13.348	0.00	17.154	0.00
180	658.618	0.09	746.131	0.09	874.702	0.08	1124.083	0.08
181	1469.549	0.21	1664.814	0.20	1951.689	0.19	2508.125	0.19

Table 6-6
Expenditure and Percentage Share of PCE,
by Sector and Year, without Productivity Growth

Scenario E

Sector	1975 MLN 1971$	%PCE	1985 MLN 1971$	%PCE	2000 MLN 1971$	%PCE	2025 MLN 1971$	%PCE
1	110.249	0.02	121.472	0.01	136.278	0.01	156.567	0.01
2	1472.691	0.21	1692.056	0.21	1941.940	0.21	2313.737	0.21
3	255.120	0.04	281.089	0.03	315.350	0.03	362.302	0.03
7	4478.586	0.63	5151.891	0.62	5912.223	0.62	7013.094	0.64
8	624.775	0.09	688.372	0.08	772.276	0.08	887.257	0.08
10	29.885	0.00	32.927	0.00	36.940	0.00	42.440	0.00
14	159.516	0.02	0.000	0.00	0.000	0.00	0.000	0.00
16	22.049	0.00	24.294	0.00	27.255	0.00	31.313	0.00
21	272.648	0.04	46.405	0.01	0.000	0.00	0.000	0.00
22	195.810	0.03	64.194	0.01	0.000	0.00	0.000	0.00
23	21955.297	3.08	24409.137	2.96	27958.801	2.95	33132.902	3.01
24	10932.098	1.53	12488.160	1.51	14150.035	1.49	16429.672	1.50
25	11014.051	1.54	12735.656	1.54	14525.602	1.53	16690.914	1.52
26	3644.708	0.51	4192.574	0.51	4776.266	0.50	5490.266	0.50
27	7892.090	1.11	9080.961	1.10	10355.676	1.09	11774.828	1.07
28	1108.353	0.16	1246.691	0.15	1400.403	0.15	1671.706	0.15
29	3096.690	0.43	3544.857	0.43	4029.162	0.43	4581.418	0.42
30	11413.430	1.60	13657.785	1.66	15848.590	1.67	17848.297	1.62
31	4843.371	0.68	5636.988	0.68	6456.207	0.68	7257.473	0.66
32	1233.298	0.17	1416.031	0.17	1600.830	0.17	1902.751	0.17
33	5798.465	0.81	6758.043	0.82	7779.594	0.82	9172.078	0.83
34	7507.699	1.05	8909.117	1.08	10524.715	1.11	12066.918	1.10
35	962.957	0.13	1151.614	0.14	1265.199	0.13	1380.327	0.13
36	1615.476	0.23	1779.919	0.22	1996.869	0.21	2294.176	0.21
37	153.902	0.02	178.275	0.02	208.156	0.02	246.523	0.02
38	3406.044	0.48	3950.613	0.48	4490.039	0.47	4970.812	0.45
39	17633.418	2.47	20598.402	2.50	23634.629	2.50	25747.422	2.34
40	2543.748	0.36	3025.327	0.37	3427.765	0.36	3722.427	0.34
41	195.816	0.03	215.749	0.03	242.046	0.03	278.083	0.03
43	335.216	0.05	384.886	0.05	439.361	0.05	513.975	0.05
45	5132.039	0.72	5654.441	0.69	6343.648	0.67	7288.133	0.66
46	256.098	0.04	282.167	0.03	316.559	0.03	363.691	0.03
48	48.300	0.01	53.216	0.01	59.703	0.01	68.592	0.01
49	2332.023	0.33	2723.522	0.33	3131.797	0.33	3544.657	0.32
51	117.921	0.02	129.924	0.02	145.760	0.02	167.462	0.02
52	2164.844	0.30	2615.617	0.32	3039.648	0.32	3586.812	0.33
53	1088.969	0.15	1377.532	0.17	1660.203	0.18	1904.198	0.17
54	1792.591	0.25	1871.204	0.23	2036.989	0.22	2159.432	0.20
55	207.216	0.03	228.310	0.03	256.137	0.03	294.273	0.03
56	102.671	0.01	115.487	0.01	126.831	0.01	136.768	0.01
57	216.910	0.03	277.781	0.03	332.994	0.04	393.927	0.04

Table 6-6 cont'd.

Sector	1975 MLN 1971$	%PCE	1985 MLN 1971$	%PCE	2000 MLN 1971$	%PCE	2025 MLN 1971$	%PCE
58	720.639	0.10	886.937	0.11	1048.740	0.11	1247.636	0.11
59	54.742	0.01	67.997	0.01	80.900	0.01	94.983	0.01
60	8.513	0.00	9.380	0.00	10.523	0.00	12.090	0.00
61	491.900	0.07	572.861	0.07	659.234	0.07	771.964	0.07
62	27.030	0.00	29.781	0.00	33.411	0.00	38.386	0.00
66	7979.000	1.12	8991.000	1.09	10255.000	1.08	12941.000	1.18
67	6673.582	0.93	7832.504	0.95	9020.551	0.95	10282.070	0.94
68	33.112	0.00	36.482	0.00	40.929	0.00	47.022	0.00
69	11063.945	1.55	12190.172	1.48	13676.000	1.44	15712.172	1.43
70	3063.590	0.43	3375.441	0.41	3786.865	0.40	4350.676	0.40
72	2637.542	0.37	3048.698	0.37	3510.500	0.37	3963.556	0.36
73	828.093	0.12	890.837	0.11	930.700	0.10	978.877	0.09
74	481.869	0.07	550.035	0.07	631.431	0.07	711.017	0.06
76	3866.048	0.54	4510.242	0.55	5137.020	0.54	5695.484	0.52
77	1161.208	0.16	1373.047	0.17	1473.126	0.16	1543.614	0.14
78	473.531	0.07	521.733	0.06	585.326	0.06	672.473	0.06
80	156.346	0.02	181.185	0.02	169.865	0.02	175.980	0.02
81	4.246	0.00	4.847	0.00	5.560	0.00	5.835	0.00
82	161.523	0.02	170.096	0.02	176.594	0.02	199.178	0.02
83	15.376	0.00	16.941	0.00	19.006	0.00	21.835	0.00
87	19.428	0.00	23.337	0.00	26.871	0.00	30.288	0.00
90	4.594	0.00	5.062	0.00	5.679	0.00	6.524	0.00
93	0.403	0.00	0.444	0.00	0.498	0.00	0.572	0.00
94	47.913	0.01	0.000	0.00	0.000	0.00	0.000	0.00
95	39.702	0.01	43.744	0.01	49.075	0.01	56.382	0.01
96	49.156	0.01	54.160	0.01	60.761	0.01	69.808	0.01
97	500.180	0.07	560.033	0.07	575.592	0.06	614.456	0.06
98	904.171	0.13	996.209	0.12	1117.635	0.12	1284.035	0.12
99	44.275	0.01	48.782	0.01	54.728	0.01	62.876	0.01
101	198.028	0.03	218.186	0.03	244.780	0.03	281.224	0.03
102	252.726	0.04	278.451	0.03	312.391	0.03	358.902	0.03
103	633.389	0.09	749.898	0.09	867.663	0.09	970.976	0.09
106	36.107	0.01	39.782	0.00	44.631	0.00	51.276	0.00
108	132.894	0.02	146.421	0.02	164.268	0.02	188.726	0.02
109	36.814	0.01	40.561	0.00	45.505	0.00	52.280	0.00
115	177.068	0.02	195.093	0.02	218.872	0.02	251.459	0.02
116	966.472	0.14	1142.561	0.14	1288.167	0.14	1421.005	0.13
117	5.175	0.00	5.702	0.00	6.397	0.00	7.349	0.00
119	15.564	0.00	17.148	0.00	19.239	0.00	22.103	0.00
120	28.421	0.00	31.314	0.00	35.131	0.00	40.361	0.00
122	1.147	0.00	1.200	0.00	1.215	0.00	1.238	0.00
123	5916.137	0.83	7130.422	0.86	8290.016	0.88	9180.430	0.84
124	850.405	0.12	1031.092	0.13	1166.577	0.12	1265.594	0.12
125	5488.293	0.77	6596.555	0.80	7640.008	0.81	8458.105	0.77
126	526.073	0.07	627.065	0.08	699.378	0.07	738.657	0.07
127	170.185	0.02	187.508	0.02	210.363	0.02	241.683	0.02

Table 6-6 cont'd.

Sector	1975 MLN 1971$	%PCE	1985 MLN 1971$	%PCE	2000 MLN 1971$	%PCE	2025 MLN 1971$	%PCE
128	287.408	0.04	324.629	0.04	364.832	0.04	390.615	0.04
129	499.515	0.07	550.362	0.07	617.445	0.07	709.374	0.06
130	175.276	0.02	202.370	0.02	233.496	0.02	264.993	0.02
131	99.056	0.01	105.657	0.01	108.523	0.01	110.788	0.01
133	33992.840	4.76	37453.062	4.54	42018.121	4.44	48274.043	4.39
134	253.601	0.04	279.415	0.03	313.472	0.03	360.144	0.03
137	508.443	0.07	560.198	0.07	628.480	0.07	722.052	0.07
139	1070.802	0.15	1173.178	0.14	1398.616	0.15	1510.554	0.14
140	1694.674	0.24	2462.409	0.30	2354.333	0.25	2577.829	0.23
142	26.881	0.00	30.599	0.00	37.631	0.00	40.538	0.00
143	463.400	0.06	518.900	0.06	581.300	0.06	725.100	0.07
144	231.700	0.03	259.450	0.03	290.650	0.03	362.550	0.03
145	1057.229	0.15	1275.070	0.15	1496.451	0.16	1623.359	0.15
146	627.456	0.09	769.307	0.09	870.531	0.09	909.749	0.08
147	2309.042	0.32	2754.036	0.33	3106.761	0.33	2948.952	0.27
148	3667.241	0.51	4295.016	0.52	4760.715	0.50	4931.934	0.45
149	297.310	0.04	327.574	0.04	367.501	0.04	422.217	0.04
150	909.256	0.13	1001.812	0.12	1123.920	0.12	1291.257	0.12
151	401.452	0.06	526.634	0.06	536.357	0.06	741.746	0.07
152	2902.484	0.41	3401.795	0.41	3992.411	0.42	4804.969	0.44
153	765.039	0.11	986.418	0.12	1132.542	0.12	1261.289	0.11
154	248.717	0.03	329.148	0.04	347.647	0.04	298.297	0.03
155	4028.213	0.56	4873.090	0.59	5528.270	0.58	6208.422	0.56
157	16.957	0.00	23.191	0.00	25.120	0.00	26.668	0.00
158	12010.980	1.68	14604.414	1.77	17011.977	1.80	19509.762	1.78
160	12160.750	1.70	13398.625	1.62	15031.750	1.59	17269.770	1.57
161	6179.148	0.87	6808.141	0.83	7637.965	0.81	8775.156	0.80
162	2935.467	0.41	3442.931	0.42	4004.799	0.42	4673.145	0.43
163	42519.375	5.96	50152.523	6.08	59255.824	6.26	71936.750	6.55
164	127017.312	17.79	149819.625	18.17	177013.813	18.70	214895.500	19.55
165	22200.590	3.11	25337.977	3.07	29036.895	3.07	30952.113	2.82
166	16804.996	2.35	19737.434	2.39	22967.836	2.43	26048.703	2.37
167	72420.500	10.15	79792.375	9.68	89518.062	9.46	102846.062	9.36
168	32742.168	4.59	43533.570	5.28	50942.965	5.38	57085.328	5.19
169	4367.750	0.61	4778.762	0.58	5366.512	0.57	5964.492	0.54
170	15918.992	2.23	19304.152	2.34	22673.777	2.40	25882.293	2.36
171	5681.480	0.80	6960.820	0.84	7935.918	0.84	8350.523	0.76
172	238.578	0.03	262.864	0.03	294.904	0.03	338.811	0.03
173	12167.488	1.70	13406.047	1.63	15040.078	1.59	17279.340	1.57
174	7699.102	1.08	9017.887	1.09	10177.367	1.08	10708.691	0.97
175	48640.898	6.81	54908.648	6.66	61485.047	6.50	76439.312	6.96
176	9115.000	1.28	9439.000	1.14	9963.000	1.05	10555.000	0.96
177	1844.597	0.26	2192.910	0.27	2573.258	0.27	3157.048	0.29
178	9.985	0.00	11.002	0.00	12.343	0.00	14.181	0.00
180	654.339	0.09	720.945	0.09	808.820	0.09	929.242	0.08
181	1460.001	0.20	1608.619	0.20	1804.689	0.19	2073.383	0.19

Table 6-7
Expenditure and Percentage Share of PCE,
by Sector and Year, without Productivity Growth

Scenario F

Sector	1975 MLN 1971$	%PCE	1985 MLN 1971$	%PCE	2000 MLN 1971$	%PCE	2025 MLN 1971$	%PCE
1	109.967	0.02	119.004	0.01	129.194	0.01	136.520	0.01
2	1472.691	0.21	1692.056	0.21	1941.940	0.21	2313.737	0.23
3	254.467	0.04	275.379	0.03	298.959	0.03	315.911	0.03
7	4478.590	0.63	5161.320	0.63	5866.914	0.63	6688.582	0.68
8	623.177	0.09	674.389	0.08	732.136	0.08	773.649	0.08
10	29.808	0.00	32.258	0.00	35.020	0.00	37.006	0.00
14	159.108	0.02	0.000	0.00	0.000	0.00	0.000	0.00
16	21.993	0.00	23.800	0.00	25.838	0.00	27.303	0.00
21	271.950	0.04	75.444	0.01	0.000	0.00	0.000	0.00
22	195.309	0.03	83.362	0.01	0.000	0.00	0.000	0.00
23	21955.297	3.08	24409.137	2.96	27958.801	3.03	33132.902	3.35
24	10932.098	1.53	12488.160	1.51	14150.035	1.53	16429.672	1.66
25	11014.051	1.54	12764.684	1.55	14418.293	1.56	15924.965	1.61
26	3644.708	0.51	4199.820	0.51	4748.355	0.51	5286.090	0.54
27	7892.094	1.11	9100.945	1.10	10279.344	1.11	11242.926	1.14
28	1108.353	0.16	1246.691	0.15	1400.403	0.15	1671.706	0.17
29	3096.690	0.43	3554.560	0.43	3998.750	0.43	4369.070	0.44
30	11413.437	1.60	13692.418	1.66	15703.801	1.70	16815.664	1.70
31	4843.371	0.68	5636.988	0.68	6456.207	0.70	7257.473	0.73
32	1233.299	0.17	1419.229	0.17	1592.531	0.17	1842.266	0.19
33	5798.469	0.81	6770.680	0.82	7727.812	0.84	8803.949	0.89
34	7507.699	1.05	8912.945	1.08	10510.098	1.14	11960.363	1.21
35	962.958	0.13	1153.095	0.14	1259.597	0.14	1338.291	0.14
36	1611.345	0.23	1743.762	0.21	1893.080	0.20	2000.418	0.20
37	153.903	0.02	178.787	0.02	207.045	0.02	238.218	0.02
38	3406.050	0.48	3966.038	0.48	4431.301	0.48	4561.535	0.46
39	17633.441	2.47	20732.602	2.51	23123.773	2.50	22387.355	2.27
40	2543.752	0.36	3048.376	0.37	3340.236	0.36	3165.994	0.32
41	195.316	0.03	211.366	0.03	229.465	0.02	242.476	0.02
43	335.217	0.05	387.772	0.05	428.159	0.05	442.193	0.04
45	5130.277	0.72	5694.234	0.69	6178.973	0.67	6138.598	0.62
46	255.443	0.04	276.435	0.03	300.106	0.03	317.122	0.03
48	48.176	0.01	52.135	0.01	56.600	0.01	59.809	0.01
49	2332.024	0.33	2728.117	0.33	3118.174	0.34	3447.757	0.35
51	117.619	0.02	127.285	0.02	138.184	0.01	146.019	0.01
52	2164.844	0.30	2615.617	0.32	3039.648	0.33	3586.812	0.36
53	1088.970	0.15	1385.729	0.17	1627.118	0.18	1677.399	0.17
54	1792.594	0.25	1882.578	0.23	1994.490	0.22	1883.545	0.19
55	206.691	0.03	223.676	0.03	242.829	0.03	256.598	0.03
56	102.671	0.01	115.702	0.01	126.364	0.01	133.906	0.01
57	216.911	0.03	278.115	0.03	331.048	0.04	380.022	0.04

Table 6-7 cont'd.

Sector	1975 MLN 1971$	%PCE	1985 MLN 1971$	%PCE	2000 MLN 1971$	%PCE	2025 MLN 1971$	%PCE
58	720.639	0.10	888.707	0.11	1042.740	0.11	1207.621	0.12
59	54.742	0.01	68.298	0.01	79.802	0.01	87.520	0.01
60	8.492	0.00	9.190	0.00	9.976	0.00	10.542	0.00
61	491.906	0.07	573.714	0.07	656.860	0.07	755.047	0.08
62	26.961	0.00	29.176	0.00	31.675	0.00	33.471	0.00
66	8012.000	1.12	9156.000	1.11	10270.000	1.11	12321.000	1.25
67	6673.582	0.94	7842.508	0.95	8979.691	0.97	9992.523	1.01
68	33.027	0.00	35.741	0.00	38.801	0.00	41.001	0.00
69	11036.566	1.55	11943.531	1.45	12966.254	1.40	13701.445	1.39
70	3055.757	0.43	3306.873	0.40	3590.039	0.39	3793.596	0.38
72	2637.545	0.37	3059.696	0.37	3468.092	0.38	3651.640	0.37
73	828.093	0.12	892.424	0.11	925.000	0.10	944.129	0.10
74	481.870	0.07	550.875	0.07	628.105	0.07	687.753	0.07
76	3866.052	0.54	4529.168	0.55	5064.270	0.55	5191.102	0.53
77	1161.211	0.16	1382.972	0.17	1436.867	0.16	1311.035	0.13
78	472.321	0.07	511.135	0.06	554.903	0.06	586.366	0.06
80	156.346	0.02	182.152	0.02	166.456	0.02	150.292	0.02
81	4.246	0.00	4.859	0.00	5.514	0.00	5.514	0.00
82	161.523	0.02	169.964	0.02	175.774	0.02	193.271	0.02
83	15.336	0.00	16.597	0.00	18.018	0.00	19.039	0.00
87	19.428	0.00	23.361	0.00	26.721	0.00	29.206	0.00
90	4.583	0.00	4.959	0.00	5.384	0.00	5.689	0.00
93	0.402	0.00	0.435	0.00	0.472	0.00	0.499	0.00
94	47.790	0.01	51.717	0.01	56.146	0.01	59.329	0.01
95	39.601	0.01	42.855	0.01	46.525	0.01	49.163	0.00
96	49.030	0.01	53.060	0.01	57.603	0.01	60.869	0.01
97	500.180	0.07	560.906	0.07	573.202	0.06	595.075	0.06
98	901.859	0.13	975.972	0.12	1059.544	0.11	1119.621	0.11
99	44.162	0.01	47.791	0.01	51.883	0.01	54.825	0.01
101	197.522	0.03	213.753	0.03	232.057	0.03	245.215	0.02
102	252.080	0.04	272.795	0.03	296.154	0.03	312.946	0.03
103	633.389	0.09	752.892	0.09	855.612	0.09	883.632	0.09
106	36.014	0.01	38.974	0.00	42.311	0.00	44.710	0.00
108	132.554	0.02	143.447	0.02	155.730	0.02	164.560	0.02
109	36.719	0.01	39.737	0.00	43.140	0.00	45.586	0.00
115	176.616	0.02	191.130	0.02	207.496	0.02	219.261	0.02
116	966.474	0.14	1146.432	0.14	1273.910	0.14	1316.479	0.13
117	5.162	0.00	5.586	0.00	6.064	0.00	6.408	0.00
119	15.524	0.00	16.800	0.00	18.239	0.00	19.273	0.00
120	28.348	0.00	30.678	0.00	33.305	0.00	35.193	0.00
122	1.147	0.00	1.205	0.00	1.195	0.00	1.083	0.00
123	5916.145	0.83	7156.848	0.87	8187.516	0.89	8443.301	0.85
124	850.406	0.12	1036.489	0.13	1143.952	0.12	1118.755	0.11
125	5488.297	0.77	6623.852	0.80	7534.441	0.82	7711.914	0.78
126	526.074	0.07	629.993	0.08	688.212	0.07	661.280	0.07
127	169.749	0.02	183.699	0.02	199.429	0.02	210.737	0.02

Table 6-7 cont´d.

Sector	1975 MLN 1971$	%PCE	1985 MLN 1971$	%PCE	2000 MLN 1971$	%PCE	2025 MLN 1971$	%PCE
128	287.409	0.04	326.680	0.04	356.896	0.04	339.750	0.03
129	498.238	0.07	539.182	0.07	585.352	0.06	618.542	0.06
130	175.276	0.02	202.939	0.02	231.318	0.03	248.382	0.03
131	99.057	0.01	106.065	0.01	106.882	0.01	98.116	0.01
133	34017.594	4.77	37850.301	4.59	40502.746	4.38	36874.285	3.73
134	252.952	0.04	273.739	0.03	297.179	0.03	314.030	0.03
137	507.143	0.07	548.819	0.07	595.814	0.06	629.597	0.06
139	1070.803	0.15	1181.057	0.14	1365.927	0.15	1274.749	0.13
140	1709.675	0.24	2485.122	0.30	2220.136	0.24	2178.653	0.22
142	26.881	0.00	30.711	0.00	37.165	0.00	37.009	0.00
143	466.400	0.07	528.800	0.06	579.900	0.06	684.100	0.07
144	233.200	0.03	264.400	0.03	289.950	0.03	342.050	0.03
145	1057.230	0.15	1282.064	0.16	1469.228	0.16	1436.183	0.15
146	627.458	0.09	775.772	0.09	846.217	0.09	758.502	0.08
147	2309.050	0.32	2791.629	0.34	2955.325	0.32	2000.760	0.20
148	3667.247	0.51	4316.492	0.52	4686.402	0.51	4448.109	0.45
149	296.550	0.04	320.920	0.04	348.400	0.04	368.155	0.04
150	906.931	0.13	981.461	0.12	1065.503	0.12	1125.918	0.11
151	401.453	0.06	528.483	0.06	530.128	0.06	692.612	0.07
152	2902.485	0.41	3411.207	0.41	3958.449	0.43	4598.918	0.47
153	765.041	0.11	992.603	0.12	1109.245	0.12	1106.918	0.11
154	248.718	0.03	333.966	0.04	328.431	0.04	180.679	0.02
155	4028.228	0.56	4930.949	0.60	5306.414	0.57	4770.480	0.48
157	16.957	0.00	23.286	0.00	24.804	0.00	24.616	0.00
158	12011.000	1.68	14650.820	1.78	16831.102	1.82	18220.961	1.84
160	12129.660	1.70	13126.449	1.59	14250.465	1.54	15058.473	1.52
161	6163.348	0.86	6669.840	0.81	7240.977	0.78	7651.543	0.77
162	2935.470	0.41	3453.430	0.42	3963.315	0.43	4376.094	0.44
163	42521.676	5.96	50227.789	6.09	57898.336	6.26	64199.715	6.50
164	127024.250	17.80	150044.500	18.19	172958.688	18.71	191782.688	19.42
165	22200.633	3.11	25515.781	3.09	28314.766	3.06	25981.258	2.63
166	16805.020	2.35	19825.941	2.40	22624.898	2.45	23668.289	2.40
167	72235.312	10.12	78171.437	9.48	84865.250	9.18	89677.125	9.08
168	32742.184	4.59	43614.035	5.29	50633.723	5.48	54962.547	5.56
169	4367.766	0.61	4839.828	0.59	5134.906	0.56	4499.273	0.46
170	15919.020	2.23	19424.320	2.36	22214.922	2.40	22874.680	2.32
171	5681.488	0.80	7006.047	0.85	7760.133	0.84	7216.691	0.73
172	237.968	0.03	257.524	0.03	279.576	0.03	295.428	0.03
173	12136.375	1.70	13133.719	1.59	14258.355	1.54	15066.812	1.53
174	7699.113	1.08	9111.258	1.10	9824.586	1.06	8467.293	0.86
175	48696.398	6.82	55725.797	6.76	60985.148	6.60	71778.812	7.27
176	9144.000	1.28	9153.000	1.11	8978.000	0.97	8465.000	0.86
177	1844.598	0.26	2196.695	0.27	2558.777	0.28	3056.058	0.31
178	9.960	0.00	10.778	0.00	11.701	0.00	12.365	0.00
180	652.665	0.09	706.300	0.09	766.781	0.08	810.257	0.08
181	1456.268	0.20	1575.942	0.19	1710.889	0.19	1807.897	0.18

in sectors 1, 2, 83, 90 and 99 (milk, eggs and poultry, steel, non-ferrous
wire and miscellaneous wire). Where spending is lower when income grows,
the decomposition analysis outlined in chapter 1 cannot be applied, and
the elasticities of change through time cannot be calculated; for such
calculations to be meaningful, some part of "growth" would have to be
attributed to changes in price or preferences. This negative effect could
then outweigh the positive effects of income and population growth. A
four-way decomposition would be possible only if the consumption functions
explicitly included the fourth factor in both the growth and static projections. Prices can be included in the INFORUM expenditure equations,
as Almon et. al. (1979) have done, but there is not yet available a complete set of projections at the sectoral level. Rather than attribute a
price or taste effect arbitrarily, we have simply suppressed the decomposition and elasticity analysis for the sectors affected.

Changes Through Time in Demand Composition

All three scenarios agree as to the direction of changes in sectoral
shares of PCE through time, assuming productivity growth (at least to an
accuracy of 0.01 percent of PCE; there are occasional discrepancies of
that size among scenarios, for sectors 56, 128, 140 and 176). The distinction between sectors which expand relative to total consumption, and
those which contract, is summarized in table 6-8. For simplicity, we consider only the entire period 1975-2025, since there are very few non-
monotonic changes in the structure of demand (compare table 1-4 and the
associated discussion). The share falls for all food sectors, tobacco,
fabrics, newspapers and books, most durable goods (furniture, pottery,
appliances, automobiles, farm machinery), rail and bus travel, insurance,
real estate and water, sewage and postal services. It rises for clothing
(but declines slightly for shoes), household textiles, periodicals,
radio and television receivers, drugs and medical goods and services,
miscellaneous leather and plastic products, water and air transportation,
telephone, both retail and wholesale trade, lodging places, personal and
repair services, business services, and amusements. There is considerable evidence of a shift away from goods and toward services, but there

Table 6-8. Direction of Change in Sectoral Shares of Total PCE,
1975-2025: All Scenarios, with and without Productivity
Growth

Nature of Change	Income Assumption	
	Growth	Static
Increase, all three scenarios	39, 40, 43, 53, 66, 74, 77, 124, 125, 143, 144, 147, 148, 153, 154, 155, 158, 163, 164, 165, 169 170, 171, 174, 175	53, 74, 97, 116, 163, 164, 168, 170
Increase in D decrease or no change in F* (share increases as population is larger)	139, 140	39, 40, 45, 123, 125, 133, 139, 140, 146, 147, 153, 154, 155, 165, 169, 171, 175.
No change or change of 0.01 percent in one scenario	56, 59, 81, 87, 122, 128, 131, 142, 157	37, 43, 56, 59, 80, 81, 82, 87, 103, 122, 126, 130, 142, 144, 145, 157
Increase in F decrease or no change in D* (share decreases as population is larger)	151, 176	2-35, 49, 52, 57, 58, 61, 66, 67, 143, 151, 152, 158, 162, 166, 175, 177
Decrease, all scenarios	2-35, 37, 38, 45, 49, 52, 54, 57, 58, 61, 67, 72, 73, 76, 80, 82, 97, 103, 116, 123, 126, 128, 130, 133, 151, 152, 162, 166, 168, 177	38, 54, 72, 73, 76, 148, 176

*Scenario E may show the same change as D, the same as F, or no changes.

are increases in some goods (clothing, all health-related goods, a few
durables such as radio, television and cycles) and decreases in certain
services (particularly those related to home-owning). For a few sectors
(usually less than 0.01 percent of PCE), there is no change over time.
The fact that all three scenarios agree so well indicates that over a long
period the income effect dominates any re-arrangement of consumption
caused by changes in population composition.

A somewhat different picture emerges if projections are made assuming
no productivity growth, so that (apart from very small income differences)
only compositional factors are at work. These results, also summarized
in table 6-8, agree with the income-growth projections for many sectors
(53, 54, 56, 59, 72, 73, 74, 76, 81, 87, 122, 131, 142, 157, 163, 164,
and 170), including most of the very small sectors. They disagree with
the other projections, or among scenarios, however, for the majority of
sectors. Food still declines as a share of PCE in scenario D, but it rises
in F, and for four sectors also in E; this reflects the greater share of
adults when population is smaller. Furniture, appliances and automobiles
rise in scenario D but fall in scenarios E and F. The same is true for
jewelry, watches, toys and sporting goods, credit services, lodging places,
business services and amusements. The opposite pattern characterizes
newspapers, drugs and medical services, telephone, water, sewage and postal services. These differences indicate that while compositional effects
are overshadowed by income effects over long periods, they may neverthe-
less be important when income differences are small.

The summary in table 6-8 concerns only the <u>direction</u> of changes in
shares. Not surprisingly, the magnitude of such changes is relatively
small if productivity is held constant, and much larger if it is assumed
to grow. For example, food sectors (2 through 33) in total absorb 13.10
percent of PCE in 1975, in scenario E; in 2025, the share is only 6.35
percent if income grows, but if there is no income growth, the share falls
only to 12.30 percent. Medical services (sector 175) provide an even more
dramatic example: if income increases, the share rises from 7.27 to 12.96
percent, whereas it actually falls slightly if productivity is held constant. Over long periods, income effects dominate those due to composition,
in any one scenario; compositional effects show up more readily when com-

paring scenarios, because of cumulated differences in population growth
rates.

When the three population scenarios are compared in a given year, the
most striking result is that there is very little difference in the composition of PCE; at least, this is true if productivity is assumed to grow.
The great majority of sectors take the same share of spending in all three
scenarios, to within 0.02 percent of PCE. There are substantial differences (of 0.05 percent or more) in only ten sectors, which, however, include some of the largest sectors. These ten in total take a roughly
constant share of PCE, implying that substitution occurs mostly within
this group:

Sector	Description	Year 2025: PCE Share in Scenario	D	E	F
39	Apparel		2.61	2.53	2.51
66	Drugs		1.99	2.19	2.34
133	Automobiles		3.59	3.54	3.52
148	Toys, sporting goods		0.70	0.64	0.63
160	Electric utilities		1.88	1.76	1.68
165	Credit agencies		3.60	3.50	3.44
168	Real Estate		4.39	4.55	4.55
174	Amusements		1.66	1.59	1.58
175	Medical services		11.98	12.96	13.61
176	Private schools		1.20	1.94	2.32
	Total, ten sectors		33.60	35.20	35.57

This comparison shows some shift away from goods and toward services as
population is smaller; since substantial parts of the shift occur in medicine and private education, composition effects appear to be more important
in producing this change than the rather slight differences in income (-2.5
percent for D compared to E, and +5.5 percent for F compared to E).

If productivity is assumed not to grow, the three scenarios differ
considerably more in PCE composition, and the direction of difference is
frequently reversed. This is particularly the case for food sectors, which
take systematically somewhat smaller shares under D than under F if productivity grows, but always take larger shares if productivity is constant.

The two projections are in agreement, however, on a decline in automobiles and a rise in health expenditures (drugs and medical services).

Average vs. Marginal Budget Shares

The finding that the composition of expenditure does not differ much among scenarios, or change appreciably through time, might be regarded as a consequence of the functional form adopted for the CES demand equations and discussed in chapter 2. This form converges to a linear equation with constant marginal budget shares; if the intercept coefficient is small, the expenditure equation eventually comes to have equal average and marginal budget shares. Beyond some income level, then, continued income growth will have no effect on the structure of spending.

This criticism is correct beyond some level of income, but it does not tell us whether the projected incomes over the next half-century come close to that level; the expenditure functions may show considerable curvature over the range in which we make projections. This can happen because the log-inverse term does not entirely die away at projected incomes. Of course, budget shares can also shift because the equations differ among household size classes, and the household size distribution differs through time and across scenarios.

Table 6-9 shows, for each of the PCE sectors estimated from the survey data, the average budget share and the asymptotic marginal budget share. These are calculated, for illustration, for a two-adult, childless household whose head is 35-44 years old, at a 1975 income of \$7,106 and a 2025 income of \$17,660. A complete comparison would require calculation for all the other combinations of household size, composition and age as well, and different income levels, but the comparison is intended only to illustrate the degree of curvature in a typical expenditure equation. Any change in the age of the household or its adult/child composition would not affect the shape of the curve, of course, but only its height; differences in curvature can occur only between household size classes. For the sectors estimated from the diary data, food at home and food away from home are first estimated as functions of income; the asymptotic marginal budget share is then of the form $\beta_{1hi}\beta_{Fhl} + \beta_{1ai}\beta_{Fal}$, where the first term in each product is a coefficient for sector i and the second term refers to total food at home or away from home.

Table 6-9. Average and Marginal Budget Shares (Percentage of Income),
by Expenditure Sector Estimated from the CES, for a Two-Adult,
Childless Household Aged 35-44

	Sectors Estimated From the Interview Data						
	Average Shares		Asymptotic		Average Shares		Asymptotic
	1975	2025	Marginal		1975	2025	Marginal
Sector	($7,106)	($17,660)	Share	Sector	($7,106)	($17,660)	Share
35	0.4167	0.2092	0.0528	128	0.1442	0.0758	0.0244
38	0.6101	0.4168	0.2770	130	0.1190	0.0510	0.0004
39	(negative)	0.4809	4.3453	131	0.0081	0.0047	0.0024
40	0.2314	0.2912	0.3321	133	(negative)	5.7906	10.9471
43	0.1190	0.0617	0.0182	139	0.1860	0.0903	0.0184
45	(negative)	0.3615	1.3576	140	(negative)	(negative)	0.4764
49	0.0269	0.0094	(negative)	142	0.0013	0.0006	0.0001
53	0.1851	0.1117	0.0559	145	0.2875	0.2410	0.2087
54	(negative)	0.1153	0.1996	146	(negative)	(negative)	0.1804
59	0.1036	0.0834	0.0662	147	(negative)	0.0529	0.5003
72	1.6501	0.8072	0.1828	148	(negative)	0.1656	0.4122
73	0.0465	0.0227	0.0057	151	0.1439	0.0744	0.0232
74	0.0061	0.0056	0.0052	152	(negative)	(negative)	0.8771
76	0.5518	0.4472	0.3705	153	(negative)	(negative)	0.4072
77	(negative)	0.0214	0.1513	154	(negative)	(negative)	0.1684
80	(negative)	0.0101	0.0585	155	(negative)	0.0090	2.4410
81	0.0024	0.0009	0.0002	157	0.0313	0.0155	0.0032
97	0.0086	0.0012	0.0132	158	4.9671	2.5046	0.7145
103	0.4207	0.1793	(negative)	162	1.4558	0.6911	0.1199
116	0.5710	0.2291	(negative)	165	1.9393	1.5989	1.3166
122	0.0001	0.0001	6.0000	166	13.1035	9.9178	7.5666
123	1.9468	1.0495	0.3925	168	(negative)	2.4002	6.5189
124	(negative)	0.0054	0.0818	169	(negative)	0.0750	3.6930
125	1.2966	0.7308	0.3127	170	39.1674	19.4759	6,7449
126	0.3006	0.1623	0.0617	174	(negative)	(negative)	3.0220

	Sectors Estimated From the Diary Data						
2	0.4192	0.2001	0.0436	32	0.4524	0.2020	0.0232
7	1.6171	0.6776	0.2114	33	1.7522	0.8476	0.2094
23	10.2749	4.5467	0.9186	34	0.8645	0.3606	0.0372
24	2.5820	1.2443	0.2981	56	0.0171	0.0081	0.0023
25	4.7198	2.2315	0.4549	57	0.0317	0.0143	0.0020
26	1.7929	0.7711	0.0600	58	0.1214	0.0609	0.0208
27	2.3551	1.2391	0.4545	61	0.1112	0.0535	0.0130
28	0.1696	0.0793	(negative)	67	1.7404	0.7690	0.1334
29	0.6070	0.2593	0.0838	82	0.0206	0.0093	0.0012
30	1.6430	0.6367	0.2695	87	0.0677	0.0282	0.0006
31	1.1992	0.5523	0.1601	170	1.3894	0.8049	0.4305

There are a few sectors, such as 40, 74 and 145, for which the budget
share changes very little as income rises, but the typical expenditure
equation shows pronounced curvature. The budget share commonly changes by
a factor of two or more between the 1975 and 2025 income levels, and in
the last projection year it is still quite different from the asymptotic
marginal share. It appears therefore that the functional form does not
force budget shares to remain constant. The only difficulty is that many
shares are so small that even a large relative change has almost no effect
on the absolute magnitude of the share.

Scale, Composition and Income Effects

The growth of expenditure in each sector is decomposed into three
sources, as described in chapter 1, and the results appear in tables 6-10
(scenario D), 6-11 (scenario E) and 6-12 (scenario F). These are comparable to table 1-5, which shows the decomposition for large spending categories. As before, we consider only the three periods beginning in 1975
and extending to each of the other benchmark years. No decomposition is
calculated for a sector if expenditure is projected to be lower with
income growth than without it. Sectors 1, 2, 8, 14, 21, 22, 69, 70, 83,
90, 94, 99, and 161 are affected. The reduction in expenditure in these
sectors is due to exogenous changes in prices or tastes, and cannot properly be attributed to any of the three factors studied. Note that for the
sectors estimated from aggregate, time-series data, the "compositional"
effect refers to all changes not explained by increases in income or
population size, although no compositional variables appear in the estimating equations. They therefore include the effects of exogenously-
imposed time trends and do not refer to specific changes in the population.

In general, the sector-by-sector results agree with those for large
categories as to the relative importance of the three sources of growth
and their changes through time and among scenarios. Income effects usually increase through time: over the entire 50-year period they account
for half or more of total growth in many sectors. Demographic effects
usually decline in importance, the change being more pronounced for composition than for scale effects. Food expenditures differ from other kinds

Table 6-10
Percentage Decomposition of Total
Expenditure Growth, by Sector
and Period: Scenario D

	Period: 1975 to								
		1985			2000			2025	
Sector	SCA	COM	INC	SCA	COM	INC	SCA	COM	INC
1	******	******	******	******	******	******	******	******	******
2	117.7	-17.7	0.0	******	******	******	******	******	******
3	37.5	-9.2	71.7	28.1	-3.3	75.2	23.2	-1.6	78.4
7	52.3	24.9	22.8	44.9	15.8	39.3	40.6	5.5	53.9
8	97.4	2.6	0.0	******	******	******	******	******	******
10	31.2	-6.4	75.2	25.8	-2.5	76.7	21.8	-1.3	79.5
14	******	******	******	******	******	******	******	******	******
16	29.0	3.4	67.6	24.1	1.3	74.5	22.1	0.7	77.2
21	******	******	******	******	******	******	******	******	******
22	******	******	******	******	******	******	******	******	******
23	66.4	-38.5	72.1	67.8	-24.5	56.7	68.2	-26.4	58.2
24	78.3	4.1	17.5	88.4	-9.8	21.4	91.7	-26.9	35.2
25	50.6	27.1	22.3	44.9	15.4	39.7	42.1	2.2	55.7
26	55.2	21.4	23.5	49.6	11.1	39.3	46.4	-1.5	55.1
27	51.3	24.4	24.3	44.3	14.0	41.7	41.0	0.6	58.5
28	85.6	-6.7	21.1	96.3	-19.7	23.4	92.5	-26.4	33.9
29	75.3	24.7	0.0	69.0	26.2	4.8	73.0	3.9	23.1
30	42.3	39.7	18.0	36.5	24.6	39.0	34.0	7.8	58.2
31	73.3	15.5	11.2	82.1	0.4	17.5	92.6	-27.8	35.2
32	67.8	30.2	2.0	67.9	13.5	18.7	66.9	0.6	32.5
33	51.0	28.9	20.1	46.0	17.5	36.5	42.6	5.1	52.3
34	63.4	30.0	6.6	64.9	19.2	15.9	70.9	-5.6	34.7
35	55.3	41.4	3.3	68.1	13.3	18.7	68.8	-13.5	44.6
36	22.2	24.8	53.0	19.0	10.1	70.9	20.9	6.6	72.5
37	56.1	24.1	19.8	50.2	16.8	33.0	48.4	4.1	47.5
38	39.4	34.2	26.4	31.5	20.7	47.8	28.4	6.8	64.8
39	29.4	40.9	29.8	20.0	25.2	54.8	16.3	11.5	72.2
40	24.8	46.8	28.5	16.3	26.1	57.6	12.7	12.4	74.9
41	58.5	23.0	18.4	42.6	8.0	49.4	38.4	4.3	57.4
43	27.3	42.6	30.1	17.8	25.6	56.6	15.1	15.2	69.6
45	27.7	11.9	60.5	25.8	19.2	55.0	21.1	13.9	64.9
46	40.1	40.5	19.5	25.2	12.1	62.7	23.7	6.8	69.5
48	58.0	-13.0	55.0	48.1	-5.1	57.0	44.0	-2.8	58.8
49	55.9	27.3	16.8	54.1	14.6	31.3	54.8	-4.1	49.3
51	31.9	-13.6	81.7	25.0	-5.0	80.0	21.2	-2.5	81.4
52	62.7	35.4	1.9	75.1	17.3	7.6	80.8	-5.8	24.9
53	26.5	57.5	16.0	19.3	36.4	44.3	16.2	19.4	64.4
54	44.2	7.5	48.3	26.2	8.8	65.0	21.6	1.7	76.7
55	8.3	-10.3	102.1	8.2	-4.9	96.7	14.4	-5.1	90.7
56	62.9	7.2	29.9	61.7	-6.6	44.9	59.6	-22.9	63.3
57	39.1	59.2	1.7	39.1	39.8	21.1	38.8	18.6	42.7

Table 6-10 cont'd.

Sector	Period: 1975 to								
	1985			2000			2025		
	SCA	COM	INC	SCA	COM	INC	SCA	COM	INC
58	43.3	47.9	8.8	41.1	30.5	28.4	40.5	13.1	46.4
59	33.1	53.0	13.9	26.3	33.8	39.9	22.6	17.3	6Q.1
60	22.3	-14.9	92.6	16.6	-5.3	88.6	14.1	-2.7	88.5
61	38.6	4.3	57.0	33.4	2.5	64.1	27.0	-3.2	76.2
62	24.2	41.5	34.2	18.5	15.0	66.5	15.9	7.7	76.4
66	20.1	1.2	78.8	14.7	0.3	85.0	10.3	1.4	88.2
67	55.4	29.2	15.4	52.9	16.0	31.1	52.2	-1.8	49.6
68	36.1	-12.6	76.5	30.2	-5.0	74.8	26.7	-2.6	75.9
69	******	******	******	******	******	******	******	******	******
70	******	******	******	******	******	******	******	******	******
72	42.6	33.1	24.3	34.9	22.1	43.0	33.9	8.3	57.8
73	55.8	-4.4	48.5	59.1	-20.4	61.3	60.6	-32.8	72.2
74	29.9	-3.1	73.2	22.9	0.1	77.0	16.5	-3.6	87.0
76	37.6	36.4	26.0	29.5	22.2	48.3	26.0	8.3	65.8
77	27.3	48.3	24.4	20.2	24.2	55.6	15.4	9.8	74.8
78	28.1	43.4	28.5	24.1	17.7	58.2	22.8	10.0	67.2
80	36.3	48.2	15.5	44.7	7.7	47.6	37.8	-3.0	65.3
81	49.2	25.1	25.7	45.7	17.1	37.2	50.5	-7.0	56.5
82	81.1	-36.7	55.6	77.2	-42.4	65.1	69.3	-36.3	67.0
83	51.5	48.5	0.0	******	******	******	******	******	******
87	49.5	42.3	8.1	50.1	24.5	25.3	50.3	2.7	47.0
90	30.2	69.8	0.0	******	******	******	******	******	******
93	1.8	-3.8	102.0	1.8	-1.8	100.0	4.7	-2.9	98.2
94	******	******	******	******	******	******	******	******	******
95	22.1	3.1	74.9	15.5	1.0	83.5	13.5	0.5	86.0
96	29.8	70.2	0.0	19.8	57.4	22.7	19.1	33.0	47.9
97	56.5	17.0	26.5	66.2	-18.2	51.9	64.6	-30.8	66.2
98	29.8	44.2	26.0	23.8	16.8	59.4	21.7	9.1	69.2
99	******	******	******	******	******	******	******	******	******
101	17.1	16.2	66.6	13.5	6.1	80.4	11.8	3.2	85.0
102	68.8	31.2	0.0	28.5	19.1	52.4	22.9	9.2	68.0
103	36.2	41.5	22.2	29.1	26.3	44.6	26.4	11.3	62.3
106	23.1	3.3	73.6	18.8	1.3	79.9	17.7	0.7	81.6
108	19.7	38.2	42.1	16.9	15.6	67.5	15.9	8.8	75.3
109	49.1	50.9	0.0	28.1	32.4	39.5	23.7	16.3	60.0
115	19.2	7.8	73.0	14.2	2.7	83.1	12.3	1.4	86.2
116	39.6	41.1	19.3	38.3	24.1	37.6	39.3	6.2	54.4
117	21.1	-8.7	87.6	14.7	-2.9	88.2	12.3	-1.4	89.2
119	26.1	60.1	13.8	15.9	17.4	66.7	14.3	9.4	76.3
120	19.1	67.9	13.1	13.1	22.2	64.6	12.2	12.3	75.6
122	46.8	2.0	51.2	38.4	-6.5	68.2	36.1	-12.0	75.9
123	36.7	45.0	18.3	30.7	28.3	40.9	28.9	10.9	60.2
124	28.9	47.1	24.1	20.8	25.8	53.4	16.5	11.0	72.5
125	28.3	31.3	40.4	21.6	19.2	59.2	17.7	6.8	75.5
126	33.6	44.2	22.2	26.7	23.5	49.9	23.6	7.4	69.0
127	14.8	21.2	64.1	11.7	8.0	80.3	10.4	4.2	85.3

Table 6-10 cont´d.   Period: 1975 to

Sector	1985			2000			2025		
	SCA	COM	INC	SCA	COM	INC	SCA	COM	INC
128	32.7	36.2	31.1	21.9	21.7	56.3	19.6	9.0	71.4
129	36.4	20.4	43.2	12.4	3.3	84.3	0.8	0.1	99.1
130	47.5	30.3	22.2	41.9	20.5	37.6	42.8	5.6	51.6
131	46.3	7.7	46.0	39.0	-3.3	64.3	37.3	-11.3	74.0
133	31.2	46.0	22.8	28.9	47.7	23.4	22.7	32.6	44.7
134	18.4	-13.1	94.7	12.9	-4.4	91.4	11.4	-2.3	90.9
137	12.9	22.1	65.0	8.6	6.9	84.5	6.9	3.3	89.7
139	33.6	24.8	41.6	19.3	21.8	58.9	16.8	11.1	72.1
140	19.0	53.4	27.5	21.9	19.3	58.7	18.9	10.4	70.7
142	47.7	28.8	23.5	33.3	28.7	38.0	36.3	9.6	54.1
143	20.4	0.1	79.5	15.1	-0.8	85.7	10.6	0.9	88.6
144	20.4	0.1	79.5	15.1	-0.8	85.7	10.6	0.9	88.6
145	29.7	47.0	23.3	21.7	29.4	48.9	19.0	13.0	68.0
146	27.0	71.7	1.3	19.6	41.2	39.3	18.2	22.3	59.5
147	24.1	75.9	0.0	14.4	51.7	34.0	14.7	37.0	48.3
148	23.5	21.9	54.6	17.3	11.3	71.4	12.7	1.5	85.8
149	30.1	12.7	57.2	23.1	4.6	72.3	21.8	2.6	75.6
150	56.9	43.1	0.0	36.6	25.2	38.2	34.1	14.0	51.8
151	28.5	62.6	8.9	36.8	26.4	36.8	27.0	22.7	50.3
152	46.3	30.5	23.2	34.3	20.2	45.4	27.9	9.5	62.6
153	24.4	59.4	16.2	20.2	35.9	44.0	16.6	17.4	66.0
154	18.1	81.9	0.0	12.8	61.4	25.9	13.1	41.4	45.4
155	17.2	51.7	31.1	10.5	29.3	60.2	7.8	16.4	75.8
157	25.4	70.4	4.2	33.6	44.0	22.4	38.4	15.9	45.7
158	29.1	28.5	42.4	23.0	17.4	59.7	17.9	6.1	75.9
160	19.4	-27.0	107.6	21.9	-14.5	92.6	23.0	-9.1	86.1
161	******	******	******	******	******	******	******	******	******
162	42.8	35.5	21.7	34.9	23.2	41.9	31.6	10.3	58.1
163	33.2	37.1	29.7	24.8	24.9	50.3	17.0	13.5	69.4
164	33.2	37.1	29.7	24.8	24.9	50.3	17.0	13.5	69.4
165	29.9	35.4	34.7	18.6	22.9	58.5	14.5	10.8	74.7
166	36.5	39.4	24.1	27.5	25.9	46.6	24.1	11.9	64.0
167	27.7	1.0	71.4	21.8	0.4	77.8	18.3	0.2	81.6
168	27.6	44.2	28.2	25.8	24.7	49.5	21.3	7.0	71.7
169	20.8	37.2	42.1	11.3	23.8	64.9	8.2	13.7	78.1
170	27.5	48.1	24.3	19.4	29.8	50.8	15.7	15.2	69.1
171	26.4	50.2	23.4	18.3	27.8	53.9	14.3	11.7	73.9
172	34.8	-12.1	77.3	21.1	-3.5	82.4	16.8	-1.7	84.9
173	48.3	13.4	38.3	35.9	4.7	59.4	30.5	2.4	67.1
174	20.3	45.5	34.2	12.5	26.1	61.4	9.5	13.0	77.5
175	19.8	1.9	78.2	14.9	-0.2	85.3	10.5	1.1	88.4
176	31.1	-4.2	73.1	19.4	-4.1	84.7	22.9	-2.8	79.9
177	52.2	36.2	11.6	47.1	23.6	29.3	43.1	11.7	45.2
178	42.0	8.5	49.4	37.7	3.6	58.7	36.1	2.1	61.8
180	26.4	-16.1	89.8	22.7	-6.6	83.9	19.1	-3.3	84.2
181	2.9	97.1	0.0	1.5	32.8	65.7	1.2	15.5	83.3

*Not calculable, or when calculated giving an attributed share
 of 10,000 percent or more. (Scale, composition and income
 effects cannot be identified because there is almost no change
 in expenditure.)

Table 6-11
Percentage Decomposition of Total
Expenditure Growth, by Sector
and Period: Scenario E

	Period: 1975 to								
	1985			2000			2025		
Sector	SCA	COM	INC	SCA	COM	INC	SCA	COM	INC
1	******	******	******	******	******	******	******	******	******
2	114.0	-14.0	0.0	******	******	******	******	******	******
3	31.4	-11.9	80.5	22.6	-4.2	81.5	17.3	-2.1	84.7
7	45.3	46.7	8.1	34.4	21.7	43.8	30.5	16.2	53.3
8	47.8	52.2	0.0	******	******	******	******	******	******
10	26.2	-8.6	82.4	20.9	-3.3	82.4	16.4	-1.7	85.3
14	******	******	******	******	******	******	******	******	******
16	22.4	1.7	75.9	18.3	0.7	81.0	15.7	0.4	83.9
21	******	******	******	******	******	******	******	******	******
22	******	******	******	******	******	******	******	******	******
23	59.7	-24.5	64.8	58.4	-5.4	47.0	55.4	2.5	42.1
24	71.5	28.5	0.0	80.2	19.8	0.0	83.5	16.5	0.0
25	43.9	48.2	7.9	35.2	22.0	42.8	32.9	13.3	53.8
26	49.6	43.8	6.5	40.5	20.8	38.7	37.2	12.4	50.3
27	45.5	46.4	8.0	35.0	20.6	44.4	32.2	11.0	56.8
28	81.6	18.4	0.0	89.6	10.4	0.0	82.7	17.3	0.0
29	71.6	28.4	0.0	56.7	37.7	5.6	63.9	24.9	11.3
30	34.9	57.1	8.0	27.2	27.2	45.6	25.4	14.3	60.3
31	62.1	37.9	0.0	70.9	29.1	0.0	84.3	15.7	0.0
32	60.4	39.6	0.0	55.3	27.8	16.8	55.1	25.0	19.9
33	43.7	49.2	7.1	35.5	24.7	39.9	32.4	17.8	49.8
34	50.4	47.8	1.7	50.5	38.4	11.0	56.7	27.4	15.9
35	41.9	53.7	4.4	47.5	23.6	28.8	54.0	7.3	38.8
36	16.3	21.8	62.0	13.8	8.9	77.2	14.3	6.0	79.8
37	47.7	45.7	6.6	39.1	27.2	33.7	36.7	20.8	42.4
38	34.2	54.5	11.3	24.0	20.3	55.7	21.8	8.8	69.3
39	25.2	60.0	14.9	14.4	18.7	66.9	11.9	7.4	80.7
40	21.1	63.7	15.3	11.7	17.9	70.4	9.3	7.0	83.7
41	45.9	19.6	34.5	32.6	6.7	60.7	27.5	3.7	68.8
43	23.2	60.8	16.0	12.9	17.8	69.4	10.9	10.6	78.5
45	23.3	25.6	51.0	18.3	9.7	72.0	15.7	5.4	79.0
46	27.4	33.0	39.7	17.6	10.3	72.1	15.5	5.8	78.6
48	49.2	-17.4	68.2	39.2	-6.7	67.5	33.6	-3.7	70.1
49	47.3	47.5	5.2	43.1	27.0	29.9	44.5	15.5	40.0
51	25.5	-15.5	90.0	19.5	-5.7	86.3	15.3	-2.9	87.6
52	48.9	51.1	0.0	58.4	41.6	0.0	64.0	36.0	0.0
53	20.2	68.1	11.7	12.8	27.4	59.8	10.9	14.9	74.2
54	43.4	35.7	21.0	20.7	4.8	74.6	17.0	-2.1	85.1
55	6.5	-10.7	104.2	6.3	-5.0	98.7	10.6	-5.5	94.9
56	62.3	31.8	5.9	55.9	7.3	36.8	56.0	-7.2	51.2
57	28.9	66.5	4.6	27.3	42.3	30.4	27.5	31.1	41.4

Table 6-11 cont´d.

Sector	Period: 1975 to								
	1985			2000			2025		
	SCA	COM	INC	SCA	COM	INC	SCA	COM	INC
58	33.7	60.9	5.4	30.5	36.8	32.7	30.1	27.9	42.1
59	25.6	64.7	9.6	18.1	29.2	52.7	15.9	18.1	66.0
60	16.6	-15.1	98.5	12.3	-5.4	93.1	9.8	-2.8	93.0
61	31.6	15.8	52.6	25.3	9.7	65.0	19.4	6.2	74.4
62	18.2	38.0	43.8	13.7	13.8	72.5	11.0	7.2	81.8
66	13.5	3.3	83.2	10.1	2.1	87.8	6.4	3.1	90.5
67	46.3	48.7	5.0	41.3	27.3	31.3	41.3	16.4	42.3
68	29.6	-15.1	85.5	24.0	-5.9	81.9	19.8	-3.2	83.4
69	******	******	******	******	******	******	******	******	******
70	******	******	******	******	******	******	******	******	******
72	36.0	53.2	10.8	26.4	23.0	50.5	25.6	12.6	61.8
73	71.9	18.6	9.6	70.1	-17.1	47.0	77.6	-33.3	55.8
74	25.2	6.5	68.3	17.9	4.5	77.7	12.3	1.1	86.6
76	32.4	55.9	11.6	22.0	20.4	57.6	19.5	8.9	71.6
77	22.4	62.4	15.1	13.8	14.6	71.7	11.2	3.6	85.2
78	22.3	41.8	35.9	18.8	17.0	64.2	16.6	9.7	73.6
80	28.6	58.3	13.2	29.0	-1.0	72.1	27.7	-9.8	82.1
81	43.2	48.0	8.8	35.2	23.5	41.3	41.9	4.2	53.9
82	110.2	-19.3	9.1	92.4	-43.3	51.0	75.9	-27.3	51.3
83	28.5	71.5	0.0	20.4	62.4	17.2	23.4	46.2	30.3
87	38.7	56.0	5.3	36.2	31.0	32.7	38.1	18.2	43.7
90	19.9	80.1	0.0	******	******	******	******	******	******
93	1.4	-3.8	102.4	1.4	-1.8	100.5	3.3	-2.9	99.6
94	******	******	******	******	******	******	******	******	******
95	15.7	1.7	82.6	11.1	0.6	88.3	9.1	0.3	90.6
96	19.6	80.4	0.0	13.5	49.9	36.6	12.2	29.1	58.7
97	54.5	38.9	6.6	72.3	-8.9	36.6	75.0	-23.5	48.5
98	23.3	41.9	34.8	18.3	15.9	65.8	15.6	8.8	75.6
99	170.4	-70.4	0.0	******	******	******	******	******	******
101	12.8	14.4	72.7	10.0	5.4	84.5	8.2	2.9	88.9
102	38.4	61.6	0.0	18.6	15.4	66.0	14.3	7.6	78.1
103	30.5	59.0	10.5	21.7	24.5	53.9	19.7	12.2	68.1
106	16.4	1.8	81.8	13.3	0.7	85.9	11.7	0.4	87.9
108	14.9	35.5	49.6	12.6	14.6	72.8	11.1	8.3	80.6
109	32.0	68.0	0.0	19.2	27.9	52.9	15.4	14.4	70.3
115	13.9	6.1	79.9	10.3	2.2	87.5	8.4	1.2	90.5
116	32.7	58.0	9.4	28.5	25.3	46.2	30.4	12.6	57.0
117	15.4	-9.1	93.7	10.7	-3.1	92.4	8.4	-1.6	93.1
119	16.3	46.3	37.4	10.6	14.5	74.8	9.0	8.0	83.0
120	12.6	55.5	31.9	9.0	19.2	71.9	7.8	10.7	81.5
122	55.9	26.2	17.9	38.0	-11.6	73.7	35.5	-19.8	84.3
123	29.8	60.7	9.5	22.5	27.2	50.3	21.4	13.5	65.1
124	24.3	63.2	12.5	15.0	19.6	65.4	12.2	7.4	80.4
125	23.1	44.1	32.8	16.0	17.9	66.1	12.9	7.4	79.7
126	27.7	60.8	11.5	20.0	20.4	59.6	18.1	6.0	75.9
127	10.5	18.2	71.3	8.3	7.0	84.7	6.9	3.8	89.3

Table 6-11 cont'd.   Period: 1975 to

Sector	1985 SCA	1985 COM	1985 INC	2000 SCA	2000 COM	2000 INC	2025 SCA	2025 COM	2025 INC
128	27.9	56.6	15.5	16.1	16.0	67.9	14.6	5.3	80.1
129	26.8	17.0	56.2	9.4	2.9	87.7	0.6	0.1	99.3
130	40.0	50.9	9.1	32.0	25.0	42.9	32.6	14.9	52.6
131	52.0	32.5	15.5	37.5	-6.2	68.7	36.2	-17.1	80.9
133	26.5	71.2	2.3	19.8	25.7	54.5	16.3	13.7	70.0
134	11.4	-11.1	99.6	8.3	-3.9	95.6	6.9	-2.1	95.2
137	9.6	19.9	70.5	6.3	6.3	87.5	4.7	3.0	92.3
139	31.2	51.1	17.8	15.1	17.6	67.3	12.6	6.4	81.0
140	15.8	67.1	17.1	19.9	19.4	60.8	17.7	7.8	74.5
142	38.4	50.1	11.5	24.4	29.2	46.4	26.4	13.5	60.1
143	13.7	2.4	83.9	10.5	0.8	88.7	6.7	2.3	91.0
144	13.7	2.4	83.9	10.5	0.8	88.7	6.7	2.3	91.0
145	24.4	63.3	12.3	15.8	24.5	59.7	14.0	10.6	75.4
146	22.5	77.5	0.0	13.5	27.6	58.9	13.3	11.9	74.7
147	21.3	78.7	0.0	10.0	28.6	61.4	10.7	10.1	79.2
148	20.5	34.4	45.1	13.4	9.9	76.7	9.9	1.0	89.1
149	23.0	10.6	66.4	17.3	3.9	78.8	15.3	2.2	82.5
150	39.3	60.7	0.0	25.8	22.0	52.2	22.7	12.5	64.9
151	22.5	70.0	7.5	25.8	22.7	51.6	19.7	27.0	53.3
152	39.4	51.4	9.2	25.3	22.8	51.9	20.2	15.6	64.2
153	19.0	69.3	11.8	13.0	25.1	61.9	11.1	12.3	76.7
154	14.1	85.9	0.0	8.1	30.0	62.0	9.2	8.6	82.2
155	14.3	67.6	18.1	7.1	17.0	75.9	5.4	7.6	87.0
157	18.2	74.2	7.5	20.8	35.7	43.5	25.6	19.7	54.8
158	23.1	39.1	37.8	16.6	17.5	65.8	12.7	8.6	78.7
160	15.2	-24.6	109.4	17.3	-13.5	96.2	17.6	-8.9	91.3
161	******	******	******	******	******	******	******	******	******
162	36.3	54.4	9.3	26.5	25.0	48.5	23.6	15.9	60.5
163	26.2	47.1	26.6	18.0	21.5	60.5	12.1	12.1	75.8
164	26.2	47.1	26.6	18.0	21.5	60.5	12.1	12.1	75.8
165	26.6	56.9	16.5	13.6	16.0	70.4	10.5	4.8	84.8
166	30.4	57.8	11.8	20.0	22.9	57.1	17.5	12.1	70.4
167	22.4	-0.6	78.2	17.2	-0.2	83.1	13.3	-0.1	86.8
168	20.5	49.7	29.7	17.9	25.9	56.2	14.6	12.1	73.3
169	18.3	58.8	22.9	7.8	12.2	80.1	5.8	5.0	89.2
170	22.7	64.1	13.2	13.6	22.8	63.6	11.2	11.8	77.0
171	21.5	65.2	13.3	12.7	20.0	67.3	10.2	6.6	83.1
172	25.4	-12.9	87.6	15.4	-3.8	88.4	11.5	-1.8	90.3
173	38.8	10.8	50.3	27.7	3.7	68.6	22.1	1.9	76.0
174	17.6	64.3	18.2	8.9	16.0	75.1	6.9	5.4	87.8
175	13.4	3.6	83.0	10.3	1.2	88.4	6.6	2.4	91.0
176	17.1	-20.4	103.4	13.4	-11.5	98.1	9.2	-7.1	97.9
177	41.4	52.8	5.8	34.4	30.5	35.1	30.7	25.8	43.5
178	33.7	6.3	60.0	29.5	2.6	67.9	26.6	1.5	71.8
180	22.5	-18.9	96.5	18.7	-7.6	88.9	14.6	-3.8	89.3
181	1.9	98.1	0.0	1.1	30.1	68.8	0.9	15.6	83.6

*Not calculable, or when calculated giving an attributed share of 10,000 percent or more. (Scale, composition and income effects cannot be identified because there is almost no change in expenditure.)

Table 6-12
Percentage Decomposition of Total
Expenditure Growth, by Sector
and Period: Scenario F

		1985		Period:	1975 to 2000			2025	
Sector	SCA	COM	INC	SCA	COM	INC	SCA	COM	INC
1	******	******	******	******	******	******	******	******	******
2	118.7	-18.7	0.0	******	******	******	******	******	******
3	26.2	-13.0	86.8	18.0	-4.5	86.6	11.7	-2.3	90.5
7	37.2	56.1	6.7	26.3	29.5	44.2	20.3	27.0	52.7
8	24.3	75.7	0.0	20.5	61.4	18.1	23.4	53.7	22.9
10	22.0	-9.5	87.5	16.8	-3.7	86.9	11.2	-1.9	90.7
14	******	******	******	******	******	******	******	******	******
16	17.3	1.0	81.7	13.8	0.4	85.8	10.1	0.2	89.7
21	******	******	******	******	******	******	******	******	******
22	******	******	******	******	******	******	******	******	******
23	52.8	-19.5	66.7	46.4	7.9	45.7	40.6	31.3	28.1
24	71.4	28.6	0.0	******	******	******	******	******	******
25	36.9	59.5	3.6	27.0	30.5	42.5	22.3	25.3	52.4
26	44.7	55.3	0.0	31.2	30.8	38.0	25.9	27.8	46.3
27	39.1	58.8	2.2	27.0	28.9	44.0	22.0	22.6	55.4
28	89.5	10.5	0.0	******	******	******	******	******	******
29	65.3	34.7	0.0	›44.3	52.2	3.5	47.1	52.1	0.8
30	27.6	63.9	8.4	20.6	33.6	45.8	16.8	22.3	60.9
31	53.1	46.9	0.0	******	******	******	******	******	******
32	53.7	46.3	0.0	43.1	42.2	14.7	38.8	50.8	10.4
33	36.8	60.8	2.4	27.1	33.4	39.5	21.6	30.5	47.9
34	39.9	55.7	4.4	38.7	52.3	9.0	******	******	******
35	40.1	59.9	0.0	36.7	39.0	24.3	35.6	30.1	34.3
36	11.9	18.9	69.3	10.0	8.1	81.9	8.8	5.5	85.7
37	37.1	50.9	12.0	30.2	36.1	33.6	25.4	36.9	37.7
38	27.6	63.2	9.2	18.2	25.4	56.4	14.5	13.1	72.4
39	19.1	63.8	17.1	10.8	20.9	68.3	7.6	7.0	85.4
40	16.3	70.9	12.8	8.6	19.8	71.6	5.8	6.1	88.1
41	34.4	16.6	48.9	24.3	6.0	69.7	17.6	3.3	79.1
43	16.1	61.7	22.2	9.3	19.5	71.2	6.9	10.0	83.2
45	17.5	33.3	49.2	13.4	13.0	73.6	9.9	3.9	86.2
46	19.0	27.1	54.0	12.4	9.0	78.6	9.2	5.1	85.7
48	40.0	-18.5	78.5	31.0	-7.3	76.4	22.9	-4.1	81.2
49	39.4	58.5	2.1	33.2	38.2	28.6	31.7	36.8	31.5
51	20.3	-15.6	95.3	15.0	-5.9	90.9	10.0	-3.0	93.0
52	40.4	59.6	0.0	******	******	******	******	******	******
53	14.1	65.4	20.4	9.4	29.1	61.5	6.8	14.8	78.4
54	29.2	33.7	37.1	16.1	6.7	77.2	11.7	-3.1	91.4
55	5.3	-10.9	105.5	4.9	-5.1	100.2	7.2	-5.7	98.5
56	73.6	26.4	0.0	46.6	22.4	31.1	43.2	16.6	40.2
57	21.7	70.8	7.4	20.1	49.6	30.4	17.5	43.7	38.8

Table 6-12 cont'd.

| Sector | Period: 1975 to | | | | | | | | |
| | 1985 | | | 2000 | | | 2025 | | |
	SCA	COM	INC	SCA	COM	INC	SCA	COM	INC
58	25.2	64.1	10.7	22.8	44.4	32.8	19.9	42.3	37.9
59	18.9	67.2	13.9	13.3	32.7	54.0	10.0	21.5	68.5
60	12.5	-14.2	101.7	9.0	-5.3	96.2	6.1	-2.7	96.6
61	26.3	22.1	51.7	19.4	15.9	64.7	12.7	14.4	72.9
62	13.7	34.3	51.9	10.1	12.9	77.0	6.9	6.8	86.3
66	9.7	7.1	83.2	6.9	4.2	88.8	3.7	4.6	91.7
67	38.4	59.5	2.1	31.7	38.1	30.2	28.8	35.7	35.6
68	23.9	-15.7	91.7	18.8	-6.3	87.5	13.2	-3.4	90.1
69	******	******	******	******	******	******	******	******	******
70	******	******	******	******	******	******	******	******	******
72	27.1	56.7	16.2	20.1	28.4	51.4	17.1	18.1	64.8
73	******	******	******	62.3	0.3	37.5	68.1	-11.7	43.6
74	21.1	10.8	68.2	13.7	8.4	77.9	8.1	5.4	86.4
76	26.2	64.9	8.9	16.6	25.0	58.4	12.7	12.2	75.2
77	19.8	80.2	0.0	10.1	17.9	72.0	6.7	3.1	90.2
78	17.3	38.8	43.9	14.4	16.5	69.2	11.0	9.6	79.4
80	38.2	61.8	0.0	22.1	9.8	68.1	15.1	-6.9	91.8
81	37.2	59.0	3.8	27.7	31.9	40.3	30.4	17.0	52.6
82	******	******	******	86.5	-32.3	45.8	63.5	-5.6	42.2
83	16.0	84.0	0.0	12.9	50.1	37.0	12.2	36.2	51.6
87	33.8	66.2	0.0	27.2	41.6	31.3	24.9	34.6	40.5
90	12.3	87.7	0.0	13.1	69.9	17.0	17.2	69.8	13.0
93	1.1	-3.7	102.6	1.0	-1.8	100.8	2.1	-2.9	100.7
94	******	******	******	******	******	******	******	******	******
95	11.4	1.1	87.6	8.0	0.4	91.6	5.6	0.2	94.2
96	12.7	87.3	0.0	9.3	43.7	47.0	7.2	25.7	67.1
97	74.7	25.3	0.0	61.5	13.7	24.8	59.9	4.7	35.4
98	17.9	38.4	43.7	13.9	15.2	71.0	10.2	8.5	81.3
99	71.9	28.1	0.0	******	******	******	******	******	******
101	9.6	12.8	77.5	7.4	5.0	87.6	5.2	2.7	92.1
102	23.1	47.3	29.6	12.4	13.0	74.6	8.1	6.4	85.5
103	24.7	69.5	5.9	16.0	29.1	55.0	12.8	15.9	71.3
106	11.7	1.2	87.1	9.5	0.5	90.0	7.1	0.3	92.7
108	11.2	31.9	56.9	9.4	13.7	76.9	7.1	7.9	85.0
109	21.3	76.7	2.0	13.3	24.4	62.3	9.0	12.6	78.4
115	10.2	5.1	84.8	7.5	1.9	90.6	5.2	1.0	93.8
116	27.8	71.8	0.4	21.7	32.8	45.5	20.0	20.3	59.8
117	11.4	-8.6	97.2	7.8	-3.0	95.2	5.2	-1.5	96.3
119	10.5	35.8	53.7	7.2	12.5	80.3	5.2	6.9	87.9
120	8.4	44.7	46.9	6.2	16.8	77.0	4.6	9.5	86.0
122	90.4	9.6	0.0	29.5	-3.2	73.7	25.7	-20.7	94.9
123	22.9	66.0	11.1	16.8	32.1	51.0	14.0	18.3	67.7
124	19.6	73.4	6.9	11.0	22.6	66.4	7.6	8.2	84.2
125	17.6	47.7	34.7	12.0	21.2	66.8	8.3	9.6	82.1
126	21.3	66.4	12.3	15.0	24.7	60.3	11.9	8.0	80.1
127	7.5	15.7	76.8	5.9	6.3	87.8	4.2	3.4	92.5

Table 6-12 cont´d. Period: 1975 to

Sector	1985			2000			2025		
	SCA	COM	INC	SCA	COM	INC	SCA	COM	INC
128	19.2	55.3	25.5	12.0	18.3	69.7	9.6	5.2	85.2
129	19.6	14.5	65.9	7.1	2.7	90.2	0.4	0.1	99.5
130	30.6	55.4	13.9	24.5	32.0	43.4	22.3	24.4	53.3
131	74.5	25.5	0.0	29.0	2.7	68.3	26.1	-16.0	89.9
133	17.6	68.3	14.1	14.5	26.9	58.6	10.4	5.3	84.3
134	7.5	-9.1	101.6	5.5	-3.4	97.9	3.8	-1.8	98.0
137	7.2	17.9	74.9	4.6	5.8	89.6	2.9	2.8	94.3
139	19.0	40.2	40.8	11.5	18.3	70.3	8.8	4.2	87.0
140	10.9	62.4	26.7	12.5	15.3	72.2	8.8	4.4	86.8
142	23.4	41.0	35.6	18.4	32.6	49.0	18.8	18.4	62.8
143	9.8	6.2	84.0	7.2	2.8	89.9	4.0	3.7	92.3
144	9.8	6.2	84.0	7.2	2.8	89.9	4.0	3.7	92.3
145	17.7	63.4	18.9	11.7	27.1	61.2	9.1	11.3	79.6
146	17.2	82.8	0.0	10.0	29.6	60.4	8.3	9.5	82.2
147	14.8	85.2	0.0	7.1	27.9	65.0	6.6	0.5	92.8
148	18.1	45.5	36.4	10.2	13.1	76.8	6.4	2.3	91.3
149	17.6	9.3	73.1	12.9	3.5	83.6	9.7	2.0	88.3
150	26.2	55.1	18.6	18.1	19.4	62.6	13.5	11.1	75.4
151	24.6	75.4	0.0	19.3	33.0	47.7	10.8	30.6	58.6
152	28.5	50.3	21.1	19.5	27.8	52.6	13.2	22.5	64.3
153	15.3	79.1	5.6	9.6	27.9	62.4	6.6	11.9	81.6
154	10.6	89.4	0.0	5.4	29.8	64.8	5.1	-1.3	96.2
155	10.6	70.5	18.9	5.2	17.2	77.6	3.2	4.5	92.2
157	14.4	78.9	6.7	16.1	42.3	41.6	15.9	25.0	59.1
158	17.8	43.4	38.8	12.4	21.2	66.3	8.1	11.9	80.0
160	13.5	-29.5	116.0	14.5	-16.2	101.6	12.4	-10.5	98.2
161	******	******	******	******	******	******	******	******	******
162	28.7	62.4	8.9	19.8	30.8	49.4	15.5	22.8	61.7
163	19.9	50.1	29.9	13.2	23.5	63.3	7.6	12.8	79.7
164	19.9	50.1	29.9	13.2	23.5	63.3	7.6	12.8	79.7
165	19.7	59.0	21.2	10.2	17.4	72.5	6.8	3.1	90.1
166	22.9	61.0	16.1	15.0	26.6	58.4	11.3	14.7	74.0
167	18.1	-1.2	83.1	13.4	-0.5	87.1	8.8	-0.2	91.4
168	15.9	53.5	30.6	13.5	31.0	55.5	9.2	18.0	72.8
169	13.2	57.9	28.9	5.9	12.0	82.1	3.6	1.5	94.9
170	17.0	66.8	16.2	10.1	25.0	64.9	7.0	12.0	81.0
171	16.0	67.5	16.5	9.4	22.0	68.6	6.5	6.2	87.4
172	18.6	-12.1	93.5	11.3	-3.7	92.5	7.2	-1.8	94.7
173	30.5	9.3	60.2	21.0	3.3	75.7	14.3	1.7	84.0
174	13.3	67.8	18.9	6.6	16.7	76.7	4.3	2.7	93.0
175	9.6	7.3	83.1	7.2	3.2	89.7	3.9	3.8	92.3
176	14.6	-25.0	110.4	11.3	-16.2	104.9	6.9	-10.6	103.7
177	31.3	57.7	11.0	25.8	38.6	35.6	19.9	39.2	40.9
178	26.5	5.1	68.5	22.6	2.2	75.2	17.6	1.3	81.1
180	19.4	-20.5	101.1	15.3	-8.2	93.0	10.1	-4.2	94.0
181	1.2	84.2	14.6	0.8	27.2	72.0	0.5	15.0	84.4

*Not calculable, or when calculated giving an attributed share of 10,000 percent or more. (Scale, composition and income effects cannot be identified because there is almost no change in expenditure.)

of spending in depending more on population than on income growth, and in
retaining relatively large composition effects (typically negative in scenario D and positive in E and F) even over long periods.

The individual sector decompositions differ from those for large
categories, described in chapter 1, chiefly in showing more variation.
For example, at the twelve-category level there are negative compositional
contributions only for energy, education and construction, but such effects
appear, in one or more scenarios, for a number of sectors in other categories such as 10, 24, 25 and 28 (foods), 48 and 51 (paper products), 55,
60 and 68 (chemicals), 82 (stone and clay), 93 (metal drums), 117 and 119
(machinery), and 172 (advertising). There is also, naturally, more variation in the size of the individual scale and income effects, as consumption within a large category shifts from some sectors to others. However,
this variation does not much affect the comparison across scenarios; scale
effects continue, typically, to be most important in scenario D, when population grows most rapidly, while composition effects are usually largest
in scenario F.

Elasticities Through Time and Across Scenarios

The elasticities of expenditure by sector with respect to income and
demographic growth are presented in tables 6-13 (scenario D), 6-14 (scenario E), and 6-15 (scenario F): these results correspond to those for the
large categories shown in table 1-6. Both elasticities tend to converge
toward 1.0, but for many sectors one or both elasticities differ markedly
from unity even over the entire 50-year projection period. Among the sectors comprising PCE, negative demographic elasticities occur only for sectors 55 (chemicals), 93 (metal drums), 176 (private education), and 160
and 180 (electric utilities), and there are few elasticities close to zero.
A large demographic elasticity--that is, a change in expenditure large
relative to the change in population, after controlling for income change--
is always associated with a large compositional effect. (The extremely
large elasticities calculated for sector 181, directly allocated imports,
simply reflect that sector's residual nature in the original INFORUM model.)

Table 6-13
Income and Demographic Elasticities
by Sector and Period: Scenario D

| Sector | Period: 1975 to | | | | | |
| | 1985 | | 2000 | | 2025 | |
	Ey	En	Ey	En	Ey	En
1	*******	*******	*******	*******	*******	*******
2	0.000	0.857	*******	*******	*******	*******
3	0.880	0.766	0.853	0.897	0.931	0.945
7	0.200	1.437	0.298	1.293	0.437	1.102
8	0.000	1.025	*******	*******	*******	*******
10	1.077	0.805	0.923	0.914	0.977	0.954
14	*******	*******	*******	*******	*******	*******
16	1.012	1.109	0.922	1.048	0.925	1.025
21	*******	*******	*******	*******	*******	*******
22	*******	*******	*******	*******	*******	*******
23	0.542	0.436	0.336	0.670	0.366	0.673
24	0.109	1.050	0.098	0.902	0.173	0.758
25	0.201	1.488	0.302	1.287	0.448	1.039
26	0.197	1.356	0.279	1.190	0.422	0.974
27	0.217	1.434	0.321	1.264	0.482	1.011
28	0.122	0.926	0.101	0.817	0.166	0.765
29	0.000	1.302	0.025	1.316	0.126	1.041
30	0.186	1.837	0.335	1.543	0.521	1.169
31	0.073	1.196	0.084	1.004	0.172	0.752
32	0.014	1.408	0.103	1.169	0.191	1.007
33	0.180	1.515	0.270	1.316	0.412	1.091
34	0.049	1.433	0.090	1.249	0.198	0.937
35	0.027	1.674	0.103	1.166	0.267	0.842
36	0.938	1.985	0.980	1.433	0.856	1.230
37	0.164	1.394	0.230	1.279	0.346	1.065
38	0.291	1.778	0.459	1.531	0.647	1.176
39	0.412	2.213	0.684	1.953	0.922	1.481
40	0.443	2.603	0.798	2.172	1.053	1.638
41	0.147	1.362	0.397	1.159	0.485	1.084
43	0.439	2.345	0.752	2.071	0.889	1.651
45	0.924	1.394	0.605	1.597	0.723	1.452
46	0.210	1.898	0.719	1.395	0.766	1.210
48	0.458	0.786	0.432	0.906	0.470	0.950
49	0.139	1.447	0.206	1.228	0.340	0.941
51	1.163	0.590	0.995	0.820	1.027	0.904
52	0.014	1.513	0.038	1.195	0.129	0.944
53	0.232	2.814	0.537	2.350	0.774	1.753
54	0.502	1.158	0.738	1.282	0.923	1.060
55	4.205	−0.272	2.442	0.441	1.459	0.703
56	0.227	1.107	0.279	0.905	0.439	0.676
57	0.018	2.309	0.167	1.790	0.336	1.339

Table 6-13 cont'd.

| Sector | Period: 1975 to | | | | | |
| | 1985 | | 2000 | | 2025 | |
	Ey	En	Ey	En	Ey	En
58	0.088	1.976	0.221	1.593	0.364	1.236
59	0.172	2.379	0.411	1.972	0.633	1.514
60	1.794	0.348	1.457	0.713	1.396	0.848
61	0.668	1.105	0.629	1.064	0.835	0.906
62	0.547	2.468	0.912	1.645	1.021	1.342
66	1.595	1.054	1.453	1.015	1.534	1.106
67	0.128	1.481	0.208	1.254	0.350	0.972
68	0.974	0.665	0.810	0.852	0.836	0.922
69	*******	*******	*******	*******	*******	*******
70	*******	*******	*******	*******	*******	*******
72	0.252	1.698	0.384	1.512	0.517	1.180
73	0.415	0.926	0.405	0.686	0.514	0.525
74	1.081	0.901	0.994	1.004	1.275	0.825
76	0.297	1.862	0.483	1.600	0.684	1.232
77	0.353	2.510	0.692	1.913	0.992	1.435
78	0.406	2.333	0.671	1.586	0.741	1.311
80	0.180	2.160	0.370	1.147	0.578	0.937
81	0.238	1.467	0.278	1.312	0.418	0.889
82	0.345	0.563	0.355	0.487	0.431	0.543
83	0.000	1.839	*******	*******	*******	*******
87	0.074	1.766	0.173	1.402	0.335	1.042
90	0.000	2.926	*******	*******	*******	*******
93	9.771	-1.332	4.861	-0.026	2.643	0.457
94	*******	*******	*******	*******	*******	*******
95	1.398	1.130	1.376	1.057	1.338	1.030
96	0.000	2.953	0.251	2.900	0.493	2.009
97	0.219	1.277	0.310	0.752	0.443	0.589
98	0.353	2.284	0.693	1.565	0.786	1.300
99	*******	*******	*******	*******	*******	*******
101	1.458	1.845	1.387	1.372	1.358	1.197
102	0.000	1.416	0.541	1.541	0.752	1.287
103	0.261	2.011	0.443	1.710	0.630	1.304
106	1.327	1.135	1.165	1.059	1.103	1.032
108	0.788	2.647	0.970	1.724	0.994	1.384
109	0.000	1.921	0.394	1.883	0.623	1.469
115	1.499	1.373	1.423	1.164	1.382	1.087
116	0.210	1.920	0.313	1.510	0.449	1.119
117	1.751	0.601	1.545	0.824	1.487	0.907
119	0.202	2.918	0.979	1.843	1.050	1.448
120	0.234	3.799	1.016	2.231	1.079	1.653
122	0.508	1.040	0.619	0.848	0.727	0.723
123	0.211	2.076	0.390	1.723	0.582	1.272
124	0.334	2.402	0.650	1.943	0.923	1.455
125	0.585	1.979	0.725	1.700	0.966	1.275
126	0.275	2.150	0.528	1.692	0.760	1.228
127	1.532	2.247	1.471	1.548	1.419	1.291

Table 6-13 cont'd.   Period: 1975 to

Sector	1985 Ey	1985 En	2000 Ey	2000 En	2025 Ey	2025 En
128	0.398	1.978	0.677	1.771	0.854	1.326
129	0.520	1.510	1.552	1.224	4.060	1.119
130	0.210	1.577	0.297	1.403	0.404	1.098
131	0.459	1.155	0.572	0.925	0.691	0.750
133	0.299	2.281	0.218	2.203	0.431	1.873
134	2.145	0.300	1.753	0.692	1.595	0.837
137	1.692	2.461	1.821	1.642	1.773	1.341
139	0.531	1.664	0.762	1.870	0.912	1.454
140	0.509	3.282	0.713	1.694	0.852	1.382
142	0.222	1.548	0.344	1.679	0.461	1.194
143	1.595	1.007	1.453	0.954	1.534	1.062
144	1.595	1.007	1.453	0.954	1.534	1.062
145	0.316	2.363	0.573	2.017	0.801	1.469
146	0.018	3.177	0.464	2.475	0.668	1.766
147	0.000	3.519	0.443	3.241	0.533	2.334
148	0.932	1.833	1.034	1.526	1.356	1.089
149	0.814	1.386	0.906	1.170	0.899	1.090
150	0.000	1.681	0.327	1.554	0.449	1.294
151	0.121	2.836	0.314	1.575	0.472	1.559
152	0.225	1.596	0.413	1.480	0.622	1.246
153	0.249	3.011	0.522	2.282	0.797	1.674
154	0.000	4.410	0.338	3.758	0.504	2.565
155	0.618	3.428	0.999	2.841	1.200	2.178
157	0.062	3.253	0.192	1.989	0.366	1.297
158	0.604	1.872	0.710	1.604	0.971	1.248
160	2.439	−0.430	1.331	0.371	1.092	0.666
161	*******	*******	*******	*******	*******	*******
162	0.223	1.744	0.372	1.536	0.536	1.237
163	0.376	1.988	0.555	1.782	0.855	1.532
164	0.376	1.988	0.555	1.782	0.855	1.532
165	0.478	2.041	0.769	1.938	1.009	1.505
166	0.282	1.957	0.480	1.735	0.676	1.348
167	1.116	1.032	1.034	1.014	1.090	1.008
168	0.406	2.378	0.533	1.746	0.835	1.237
169	0.759	2.527	1.081	2.473	1.253	1.981
170	0.349	2.495	0.631	2.133	0.869	1.631
171	0.345	2.613	0.699	2.126	0.995	1.545
172	1.017	0.666	1.145	0.853	1.207	0.922
173	0.365	1.256	0.549	1.113	0.659	1.060
174	0.615	2.867	0.966	2.466	1.201	1.844
175	1.595	1.091	1.453	0.986	1.534	1.077
176	1.047	0.871	1.250	0.810	0.964	0.904
177	0.101	1.627	0.211	1.411	0.344	1.199
178	0.535	1.188	0.525	1.083	0.548	1.044
180	1.514	0.403	1.126	0.738	1.133	0.861
181	0.000	13.781	1.627	7.537	1.786	4.470

*Not calculable, or estimated as 1,000 or higher because of low expenditure growth.

Table 6-14
Income and Demographic Elasticities
by Sector and Period: Scenario E

Sector	Period: 1975 to					
	1985		2000		2025	
	Ey	En	Ey	En	Ey	En
1	*******	*******	*******	*******	*******	*******
2	0.000	0.882	*******	*******	*******	*******
3	0.824	0.631	0.802	0.831	0.888	0.898
7	0.056	1.939	0.293	1.536	0.358	1.415
8	0.000	1.992	*******	*******	*******	*******
10	0.984	0.682	0.856	0.854	0.923	0.912
14	*******	*******	*******	*******	*******	*******
16	1.014	1.073	0.907	1.033	0.915	1.020
21	*******	*******	*******	*******	*******	*******
22	*******	*******	*******	*******	*******	*******
23	0.372	0.602	0.217	0.916	0.194	1.037
24	0.000	1.373	0.000	1.217	0.000	1.161
25	0.056	1.995	0.281	1.533	0.349	1.323
26	0.042	1.809	0.229	1.442	0.301	1.269
27	0.056	1.929	0.294	1.502	0.377	1.274
28	0.000	1.213	0.000	1.103	0.000	1.172
29	0.000	1.372	0.025	1.563	0.044	1.311
30	0.068	2.452	0.355	1.824	0.457	1.441
31	0.000	1.565	0.000	1.356	0.000	1.153
32	0.000	1.606	0.077	1.433	0.087	1.358
33	0.050	2.022	0.259	1.589	0.320	1.430
34	0.011	1.866	0.053	1.640	0.068	1.381
35	0.032	2.152	0.150	1.428	0.181	1.111
36	1.021	2.201	0.995	1.549	0.878	1.332
37	0.044	1.875	0.203	1.589	0.249	1.442
38	0.098	2.416	0.482	1.709	0.591	1.323
39	0.164	3.051	0.800	2.044	0.963	1.479
40	0.191	3.540	0.948	2.208	1.121	1.570
41	0.242	1.398	0.439	1.182	0.522	1.110
43	0.188	3.236	0.889	2.105	0.947	1.719
45	0.632	1.998	0.775	1.456	0.831	1.276
46	0.427	2.089	0.794	1.498	0.828	1.301
48	0.467	0.657	0.437	0.843	0.479	0.905
49	0.034	1.914	0.167	1.534	0.210	1.280
51	1.113	0.405	0.954	0.728	0.996	0.836
52	0.000	1.951	0.000	1.601	0.000	1.440
53	0.150	3.793	0.731	2.615	0.864	1.971
54	0.153	1.755	0.756	1.203	0.904	0.894
55	3.787	-0.704	2.252	0.221	1.374	0.529
56	0.031	1.475	0.172	1.117	0.240	0.890
57	0.045	2.992	0.226	2.224	0.279	1.823

Table 6-14 cont'd.

	Period: 1975 to					
	1985		2000		2025	
Sector	Ey	En	Ey	En	Ey	En
58	0.048	2.595	0.230	1.978	0.273	1.690
59	0.104	3.163	0.528	2.266	0.633	1.827
60	1.760	0.093	1.415	0.585	1.365	0.749
61	0.515	1.467	0.560	1.333	0.691	1.256
62	0.642	2.821	0.915	1.833	1.013	1.503
66	1.662	1.232	1.422	1.184	1.494	1.379
67	0.034	1.957	0.180	1.561	0.233	1.317
68	0.927	0.501	0.776	0.772	0.811	0.862
69	*******	*******	*******	*******	*******	*******
70	*******	*******	*******	*******	*******	*******
72	0.090	2.318	0.406	1.726	0.471	1.389
73	0.045	1.243	0.188	0.775	0.212	0.612
74	0.820	1.243	0.870	1.220	1.078	1.075
76	0.106	2.525	0.528	1.769	0.652	1.363
77	0.182	3.360	0.898	1.869	1.085	1.262
78	0.447	2.648	0.660	1.754	0.728	1.456
80	0.132	2.781	0.583	0.968	0.662	0.686
81	0.064	2.007	0.270	1.565	0.306	1.084
82	0.029	0.831	0.163	0.557	0.196	0.680
83	0.000	3.150	0.144	3.169	0.210	2.310
87	0.041	2.295	0.206	1.715	0.253	1.377
90	0.000	4.266	*******	*******	*******	*******
93	8.764	-2.057	4.465	-0.399	2.495	0.154
94	*******	*******	*******	*******	*******	*******
95	1.472	1.100	1.366	1.045	1.326	1.028
96	0.000	4.316	0.394	3.513	0.588	2.519
97	0.039	1.659	0.141	0.888	0.186	0.723
98	0.419	2.585	0.691	1.725	0.778	1.438
99	0.000	0.599	*******	*******	*******	*******
101	1.457	2.021	1.343	1.467	1.322	1.282
102	0.000	2.424	0.686	1.693	0.849	1.419
103	0.100	2.697	0.492	1.922	0.605	1.482
106	1.409	1.106	1.188	1.049	1.127	1.029
108	0.844	3.053	0.959	1.939	0.982	1.567
109	0.000	2.849	0.513	2.152	0.701	1.696
115	1.546	1.412	1.401	1.189	1.363	1.114
116	0.084	2.565	0.350	1.739	0.389	1.331
117	1.753	0.419	1.503	0.734	1.452	0.839
119	0.582	3.401	1.083	2.098	1.117	1.664
120	0.578	4.535	1.097	2.617	1.129	1.977
122	0.105	1.437	0.494	0.716	0.592	0.486
123	0.092	2.777	0.444	1.981	0.548	1.488
124	0.141	3.217	0.760	2.054	0.948	1.473
125	0.396	2.676	0.748	1.917	0.912	1.448
126	0.117	2.901	0.582	1.842	0.735	1.269
127	1.625	2.538	1.468	1.703	1.414	1.425

Table 6-14 cont'd.   Period: 1975 to

Sector	1985		2000		2025	
	Ey	En	Ey	En	Ey	En
128	0.158	2.769	0.770	1.817	0.876	1.293
129	0.630	1.589	1.472	1.269	3.781	1.163
130	0.070	2.146	0.300	1.656	0.341	1.362
131	0.096	1.578	0.460	0.848	0.554	0.571
133	0.024	3.285	0.523	2.045	0.680	1.631
134	2.378	0.032	1.826	0.557	1.653	0.733
137	1.693	2.813	1.753	1.829	1.721	1.501
139	0.168	2.454	0.788	1.952	0.947	1.399
140	0.259	4.409	0.596	1.805	0.722	1.349
142	0.091	2.173	0.386	1.970	0.447	1.403
143	1.662	1.167	1.423	1.069	1.494	1.276
144	1.662	1.167	1.423	1.069	1.494	1.276
145	0.138	3.214	0.661	2.226	0.810	1.577
146	0.000	3.844	0.700	2.554	0.813	1.669
147	0.000	4.036	0.843	3.061	0.969	1.701
148	0.608	2.489	1.000	1.624	1.238	1.082
149	0.855	1.430	0.903	1.197	0.900	1.119
150	0.000	2.377	0.428	1.710	0.531	1.429
151	0.089	3.603	0.422	1.734	0.437	1.972
152	0.071	2.175	0.430	1.750	0.552	1.586
153	0.158	4.001	0.760	2.482	0.905	1.807
154	0.000	5.615	0.923	3.529	1.091	1.696
155	0.295	4.745	1.303	2.765	1.431	2.002
157	0.102	4.294	0.392	2.341	0.404	1.585
158	0.460	2.501	0.727	1.864	0.899	1.517
160	2.161	−0.666	1.206	0.238	1.001	0.539
161	*******	*******	*******	*******	*******	*******
162	0.077	2.339	0.387	1.783	0.476	1.517
163	0.290	2.583	0.627	1.968	0.863	1.742
164	0.290	2.583	0.627	1.968	0.863	1.742
165	0.176	2.862	0.881	1.952	1.101	1.362
166	0.113	2.667	0.551	1.935	0.666	1.528
167	1.053	0.975	0.977	0.988	1.043	0.993
168	0.389	3.082	0.573	2.151	0.761	1.625
169	0.320	3.684	1.373	2.232	1.487	1.650
170	0.157	3.391	0.769	2.310	0.909	1.773
171	0.164	3.547	0.857	2.241	1.071	1.499
172	1.083	0.503	1.139	0.773	1.196	0.863
173	0.414	1.263	0.565	1.120	0.669	1.073
174	0.258	3.998	1.177	2.399	1.362	1.595
175	1.662	1.250	1.422	1.106	1.494	1.289
176	1.832	−0.211	1.461	0.158	1.574	0.260
177	0.043	2.149	0.229	1.737	0.281	1.633
178	0.562	1.176	0.536	1.080	0.562	1.049
180	1.341	0.164	1.022	0.618	1.054	0.769
181	0.000	19.156	1.598	9.720	1.659	6.270

*Not calculable, or estimated as 1,000 or higher because of low expenditure growth.

Table 6-15
Income and Demographic Elasticities
by Sector and Period: Scenario F

	Period: 1975 to					
	1985		2000		2025	
Sector	Ey	En	Ey	En	Ey	En
1	*******	*******	*******	*******	*******	*******
2	0.000	0.848	*******	*******	*******	*******
3	0.705	0.515	0.762	0.762	0.852	0.823
7	0.037	2.373	0.266	1.958	0.305	2.065
8	0.000	3.685	0.119	3.287	0.112	2.704
10	0.826	0.577	0.805	0.793	0.877	0.846
14	*******	*******	*******	*******	*******	*******
16	0.931	1.057	0.895	1.028	0.906	1.021
21	*******	*******	*******	*******	*******	*******
22	*******	*******	*******	*******	*******	*******
23	0.286	0.640	0.183	1.156	0.101	1.646
24	0.000	1.380	*******	*******	*******	*******
25	0.020	2.463	0.250	1.963	0.288	1.922
26	0.000	2.137	0.201	1.852	0.229	1.876
27	0.011	2.369	0.260	1.917	0.311	1.842
28	0.000	1.112	*******	*******	*******	*******
29	0.000	1.501	0.014	2.004	0.002	1.902
30	0.059	3.049	0.323	2.345	0.403	2.063
31	0.000	1.823	*******	*******	*******	*******
32	0.000	1.803	0.059	1.845	0.037	2.048
33	0.013	2.494	0.230	2.043	0.264	2.125
34	0.023	2.276	0.039	2.139	*******	*******
35	0.000	2.360	0.112	1.913	0.135	1.705
36	1.013	2.442	1.007	1.707	0.895	1.526
37	0.067	2.254	0.180	2.016	0.184	2.150
38	0.065	3.026	0.443	2.172	0.547	1.752
39	0.160	3.857	0.751	2.568	0.952	1.763
40	0.134	4.623	0.898	2.833	1.121	1.863
41	0.301	1.456	0.472	1.224	0.554	1.167
43	0.236	4.225	0.856	2.686	0.945	2.150
45	0.521	2.709	0.739	1.835	0.855	1.337
46	0.542	2.304	0.845	1.639	0.879	1.476
48	0.437	0.547	0.442	0.778	0.489	0.835
49	0.011	2.351	0.142	1.981	0.133	1.941
51	0.978	0.238	0.923	0.626	0.970	0.722
52	0.000	2.345	*******	*******	*******	*******
53	0.238	4.816	0.701	3.357	0.859	2.642
54	0.259	2.065	0.705	1.371	0.867	0.756
55	3.104	-1.125	2.099	-0.041	1.291	0.224
56	0.000	1.340	0.122	1.428	0.141	1.332
57	0.063	3.799	0.209	2.943	0.231	2.831

Table 6-15 cont'd.

	Period:		1975 to			
	1985		2000		2025	
Sector	Ey	En	Ey	En	Ey	En
58	0.081	3.237	0.210	2.579	0.211	2.600
59	0.130	4.025	0.503	2.939	0.600	2.616
60	1.575	-0.145	1.380	0.438	1.338	0.582
61	0.400	1.783	0.503	1.714	0.589	1.921
62	0.658	3.203	0.916	2.080	1.008	1.805
66	1.465	1.690	1.387	1.541	1.429	1.990
67	0.012	2.407	0.156	2.020	0.160	1.999
68	0.812	0.355	0.751	0.684	0.791	0.764
69	*******	*******	*******	*******	*******	*******
70	*******	*******	*******	*******	*******	*******
72	0.117	2.868	0.375	2.183	0.432	1.865
73	*******	*******	0.117	1.004	0.110	0.843
74	0.646	1.481	0.788	1.540	0.944	1.567
76	0.065	3.179	0.488	2.251	0.615	1.791
77	0.000	4.396	0.833	2.446	1.113	1.399
78	0.463	2.991	0.654	1.976	0.714	1.727
80	0.000	2.468	0.491	1.398	0.755	0.569
81	0.021	2.441	0.232	1.981	0.240	1.476
82	*******	*******	0.110	0.645	0.112	0.919
83	0.000	5.238	0.328	3.837	0.381	3.113
87	0.000	2.757	0.179	2.273	0.202	2.105
90	0.000	6.467	0.140	4.618	0.072	3.696
93	7.164	-2.785	4.148	-0.858	2.346	-0.382
94	*******	*******	*******	*******	*******	*******
95	1.393	1.090	1.356	1.044	1.316	1.033
96	0.000	6.313	0.502	4.290	0.661	3.451
97	0.000	1.322	0.078	1.203	0.097	1.069
98	0.450	2.914	0.691	1.938	0.770	1.699
99	0.000	1.369	*******	*******	*******	*******
101	1.342	2.221	1.309	1.598	1.290	1.446
102	0.246	2.828	0.786	1.896	0.928	1.668
103	0.045	3.454	0.464	2.486	0.569	2.001
106	1.353	1.098	1.203	1.048	1.144	1.036
108	0.840	3.488	0.952	2.219	0.970	1.908
109	0.017	4.059	0.599	2.499	0.767	2.117
115	1.450	1.474	1.383	1.232	1.345	1.173
116	0.003	3.269	0.310	2.256	0.359	1.833
117	1.591	0.254	1.469	0.634	1.424	0.728
119	0.819	3.915	1.154	2.429	1.167	2.064
120	0.804	5.307	1.154	3.111	1.165	2.573
122	0.000	1.102	0.440	0.898	0.558	0.214
123	0.091	3.509	0.413	2.555	0.507	2.045
124	0.063	4.164	0.717	2.653	0.924	1.874
125	0.357	3.370	0.692	2.451	0.853	1.934
126	0.106	3.689	0.537	2.357	0.699	1.571
127	1.556	2.855	1.463	1.909	1.408	1.677

Table 6-15 cont'd. Period: 1975 to

Sector	1985		2000		2025	
	Ey	En	Ey	En	Ey	En
128	0.242	3.504	0.731	2.267	0.852	1.461
129	0.659	1.691	1.409	1.338	3.535	1.252
130	0.091	2.629	0.273	2.101	0.295	1.894
131	0.000	1.326	0.407	1.085	0.513	0.413
133	0.140	4.267	0.529	2.512	0.806	1.437
134	2.293	-0.220	1.864	0.402	1.694	0.554
137	1.550	3.193	1.698	2.075	1.676	1.801
139	0.400	2.890	0.757	2.319	0.921	1.408
140	0.372	5.585	0.750	2.037	0.919	1.431
142	0.296	2.584	0.373	2.450	0.395	1.806
143	1.465	1.590	1.387	1.352	1.429	1.771
144	1.465	1.590	1.387	1.352	1.429	1.771
145	0.189	4.054	0.626	2.833	0.784	2.005
146	0.000	4.940	0.666	3.272	0:859	1.932
147	0.000	5.583	0.850	3.855	1.180	1.069
148	0.368	3.219	0.911	2.088	1.158	1.314
149	0.804	1.495	0.899	1.243	0.901	1.181
150	0.138	2.874	0.504	1.918	0.598	1.684
151	0.000	3.650	0.352	2.405	0.470	3.028
152	0.147	2.591	0.392	2.193	0.487	2.327
153	0.060	5.180	0.707	3.228	0.929	2.396
154	0.000	7.279	0.939	4.733	1.426	0.763
155	0.262	6.184	1.248	3.486	1.520	2.105
157	0.076	5.413	0.334	3.045	0.400	2.233
158	0.397	3.151	0.672	2.405	0.828	2.159
160	1.755	-1.292	1.110	-0.123	0.948	0.163
161	*******	*******	*******	*******	*******	*******
162	0.061	2.935	0.362	2.293	0.427	2.167
163	0.278	3.213	0.612	2.457	0.844	2.308
164	0.278	3.213	0.612	2.457	0.844	2.308
165	0.196	3.592	0.841	2.407	1.102	1.395
166	0.133	3.337	0.517	2.455	0.635	2.038
167	0.915	0.936	0.933	0.969	1.003	0.977
168	0.334	3.878	0.516	2.827	0.678	2.485
169	0.358	4.645	1.308	2.644	1.582	1.359
170	0:165	4.307	0.725	2.944	0.896	2.329
171	0.175	4.518	0.808	2.849	1.060	1.787
172	1.028	0.357	1.132	0.685	1.187	0.765
173	0.416	1.290	0.577	1.142	0.681	1.106
174	0.228	5.152	1.117	2.993	1.415	1.542
175	1.465	1.707	1.387	1.396	1.429	1.794
176	1.545	-0.765	1.392	-0.495	1.486	-0.641
177	0.070	2.659	0.213	2.250	0.230	2.497
178	0.538	1.183	0.546	1.090	0.575	1.067
180	1.088	-0.058	0.940	0.481	0.984	0.614
181	0.416	24.371	1.607	12.455	1.581	9.534

*Not calculable, or estimated as 1,000 or higher because of low expenditure growth.

The income elasticities are never negative (calculations which would show a negative value have been suppressed, as explained above), but they are often zero or very low. As is to be expected, the income elasticity is always below 1.0 for food sectors; it is extremely high--often over 2.0--for health sectors (66, 143, 144 and 175). It is somewhat misleading to call this an "income" effect, since it comes about largely from a shift from private to public or third-party payment for medical care, which is a price effect for the consumer. Private education also shows a high elasticity, due in part to increases in unit costs. High elasticities also characterize air and water transport, and hotels and lodging places, and these are truly income effects.

Demographic elasticities can also be computed between any two scenarios: we show comparisons of D to E and F to E in table 6-16, for each of the three future benchmark years. These calculations are based on the income-growth scenarios alone, so there is no difficulty in separating income and demographic effects. Calculations are suppressed only for sectors in which expenditure drops to zero. This table is comparable to table 1-7, which shows the cross-scenario elasticities for large categories. An elasticity of 1.0 means that expenditure differs between the two scenarios in the same proportion as population, so that population composition and income differences together have no net effect on spending. A negative elasticity means that expenditure and population differ in opposite directions, spending being lower in scenario D despite the higher population or larger in scenario F despite the smaller population. Negative elasticities indicate a combined income and composition effect large enough to more than offset the population size difference. These are concentrated in scenario F in 1985, become less common in the year 2000 and almost disappear by 2025: by then the population difference between scenarios E and F is too large to be offset readily. In all years, negative elasticities are less common in comparisons of scenarios D and E. Income and population composition both differ slightly more from that of scenario E in scenario F than they do in scenario D; scenario E, in this sense, is not midway between the other two alternatives.

By the end of the projection period, most of the cross-scenario elasticities have become fairly close to 1.0, but there are still a number

Table 6-16. Demographic Elasticities, $e_N(E)$, Scenarios D and
F Compared to E, by Sector and Year (Assuming
Income Growth)

Year:	1985		2000		2025	
Sector Scenario:	D	F	D	F	D	F
1	0.000	0.000	0.000	0.000	0.000	0.000
2	0.691	0.683	0.680	0.251	0.789	0.638
3	0.659	0.398	0.760	0.662	0.882	0.820
7	0.824	0.038	0.393	0.191	0.775	0.579
8	-1.899	-4.772	-1.179	-1.768	-0.636	-0.803
10	0.767	0.543	0.837	0.766	0.921	0.881
14	*******	*******	*******	*******	*******	*******
16	0.092	-0.740	0.399	0.162	0.708	0.582
21	-22.712	-23.680	*******	*******	*******	*******
22	-8.853	-12.731	*******	*******	*******	*******
23	0.839	0.246	0.753	0.223	0.793	0.652
24	0.749	0.971	0.763	0.312	0.874	0.695
25	0.843	0.298	0.502	0.249	0.873	0.645
26	0.952	0.794	0.655	0.243	0.912	0.696
27	0.956	0.431	0.548	0.278	0.905	0.667
28	0.840	1.251	0.783	0.317	0.834	0.613
29	0.972	0.659	0.572	0.269	0.923	0.660
30	0.609	-0.182	0.268	0.213	0.797	0.624
31	0.504	0.138	0.656	0.279	0.867	0.800
32	0.771	0.587	0.538	0.253	0.828	0.595
33	0.746	0.351	0.397	0.228	0.790	0.577
34	0.228	-0.248	0.349	0.163	0.727	0.690
35	-0.129	1.987	-0.156	0.435	0.673	0.495
36	-0.482	-1.959	0.088	-0.255	0.557	0.398
37	0.731	-0.642	0.459	0.186	0.756	0.643
38	1.048	0.048	0.434	0.256	0.953	0.668
39	1.241	-0.755	0.117	0.198	0.910	0.596
40	1.187	-0.024	0.047	0.122	0.970	0.586
41	0.128	-0.673	0.328	0.069	0.617	0.462
43	1.192	-1.715	0.142	-0.039	0.863	0.610
45	1.066	-0.482	-0.006	0.130	0.909	0.616
46	-0.869	-2.138	-0.209	-0.579	0.377	0.148
48	0.523	0.072	0.612	0.451	0.771	0.668
49	0.630	0.191	0.509	0.237	0.876	0.710
51	0.355	-0.201	0.556	0.378	0.781	0.679
52	0.096	0.160	0.298	0.217	0.649	0.619
53	0.143	-2.156	-0.565	-0.021	0.545	0.505
54	1.967	-2.458	0.840	0.266	1.168	0.813
55	0.515	0.483	0.681	0.631	0.909	0.937
56	1.274	2.080	0.990	0.576	1.185	0.825
57	-0.330	-0.473	-0.197	0.096	0.574	0.504

Table 6-16 cont'd.

Sector	Year: Scenario:	1985 D	1985 F	2000 D	2000 F	2025 D	2025 F
58		0.149	-0.775	0.196	0.121	0.724	0.592
59		0.181	-0.951	-0.284	0.019	0.666	0.524
60		-0.332	-1.365	0.208	-0.077	0.644	0.488
61		0.556	0.213	0.364	0.240	0.712	0.613
62		-0.219	-1.208	0.249	-0.027	0.654	0.504
66		-1.471	-2.756	-0.375	-1.099	0.222	0.086
67		0.546	0.195	0.426	0.219	0.820	0.670
68		0.504	0.035	0.643	0.494	0.817	0.733
69		0.033	0.060	0.510	0.440	0.805	0.867
70		0.996	1.003	1.013	1.017	1.114	0.955
72		0.857	-0.875	0.401	0.207	0.848	0.660
73		2.415	5.922	1.954	0.868	1.797	0.972
74		0.986	0.239	0.671	0.273	0.951	0.747
76		1.018	0.121	0.308	0.238	0.892	0.626
77		0.662	2.438	-0.433	0.362	0.831	0.389
78		0.319	-0.605	0.564	0.360	0.792	0.720
80		-0.052	7.326	-0.696	0.999	0.651	0.048
81		1.016	0.367	0.452	0.380	0.987	0.777
82		2.037	2.214	1.698	0.637	1.414	0.791
83		-2.210	-5.280	-1.095	-1.675	-0.275	-0.488
87		0.089	1.179	0.006	0.246	0.699	0.542
90		-1.411	-4.738	-0.646	-1.251	-0.173	-0.284
93		0.127	0.085	0.511	0.441	0.805	0.866
94		*******	*******	*******	*******	*******	*******
95		-0.816	-2.185	-0.011	-0.353	0.559	0.376
96		-1.454	-3.907	-0.425	-0.903	0.300	0.099
97		1.125	3.734	1.514	0.787	1.539	0.847
98		0.179	-0.764	0.470	0.243	0.746	0.649
99		-0.654	-2.442	-0.707	-1.240	-1.557	-1.475
101		-0.288	-1.575	0.277	-0.007	0.684	0.559
102		-1.700	-3.381	-0.673	-1.121	0.190	-0.086
103		0.818	0.387	0.272	0.101	0.850	0.655
106		-0.813	-2.363	-0.125	-0.508	0.457	0.258
108		-0.130	-1.495	0.324	0.037	0.688	0.578
109		-1.063	-2.440	-0.338	-0.738	0.341	0.094
115		-0.689	-2.147	0.055	-0.284	0.583	0.417
116		0.621	0.882	0.241	0.361	0.860	0.593
117		-0.569	-1.705	0.129	-0.174	0.625	0.459
119		-2.132	-4.552	-0.670	-1.153	0.241	0.002
120		-1.883	-4.368	-0.505	-0.977	0.325	0.111
122		2.487	4.376	1.752	0.516	1.608	0.935
123		0.559	-0.437	0.162	0.161	0.814	0.648
124		0.921	0.565	0.051	0.126	0.897	0.590
125		0.742	-0.765	0.269	0.206	0.863	0.664
126		0.752	-0.406	0.330	0.272	0.985	0.704
127		-1.076	-2.815	-0.149	-0.537	0.489	0.297

Table 6-16 cont´d.

Sector	Year: Scenario:	1985 D	1985 F	2000 D	2000 F	2025 D	2025 F
128		1.196	-2.120	0.306	0.094	0.959	0.750
129		-0.318	-1.366	0.453	0.242	0.980	0.969
130		0.761	-0.684	0.421	0.209	0.799	0.662
131		2.216	4.103	1.592	0.504	1.546	0.917
133		1.340	-2.441	-0.144	0.045	0.826	0.586
134		-2.714	-4.839	-1.028	-1.511	0.081	-0.209
137		-0.510	-1.832	0.137	-0.179	0.609	0.451
139		2.020	-4.556	0.913	-0.002	1.068	0.943
140		0.469	-3.041	1.467	-1.647	1.689	-0.746
142		0.639	-3.735	0.260	-0.016	0.728	0.843
143		-1.486	-2.788	-0.297	-1.025	0.283	0.153
144		-1.486	-2.788	-0.297	-1.025	0.283	0.153
145		0.835	-1.499	0.167	0.099	0.908	0.677
146		1.025	-0.158	-0.249	0.233	0.884	0.523
147		1.764	-1.329	-0.157	-0.092	0.926	0.617
148		1.459	1.763	0.662	0.419	1.218	0.735
149		-0.036	-0.772	0.309	0.067	0.656	0.507
150		-0.602	-2.044	-0.127	-0.515	0.386	0.175
151		-0.028	5.666	-0.301	0.709	0.702	0.006
152		0.855	-1.566	0.239	0.209	0.760	0.544
153		0.179	0.786	-0.883	0.286	0.493	0.279
154		0.139	1.999	-1.230	-0.160	0.707	0.378
155		1.374	-0.945	-0.434	0.087	0.785	0.450
157		-0.578	-0.080	-0.952	0.601	0.398	0.404
158		0.419	-0.532	0.104	0.156	0.751	0.592
160		0.992	0.964	1.004	0.992	1.083	0.890
161		0.996	1.003	1.013	1.017	1.114	0.955
162		0.789	-0.120	0.365	0.116	0.819	0.620
163		0.416	-0.758	0.137	-0.056	0.755	0.522
164		0.416	-0.758	0.137	-0.056	0.755	0.522
165		1.579	-1.195	0.304	0.056	0.894	0.684
166		0.843	-0.877	0.146	0.147	0.801	0.597
167		0.495	0.018	0.665	0.525	0.842	0.767
168		-0.318	-0.398	-0.278	0.273	0.565	0.539
169		2.095	-2.200	-0.199	0.218	0.888	0.502
170		0.874	-0.912	-0.119	0.135	0.804	0.522
171		0.776	-0.983	-0.182	0.143	0.821	0.590
172		-0.414	-1.395	0.138	-0.158	0.607	0.432
173		0.296	-0.291	0.422	0.219	0.697	0.547
174		1.717	-0.801	0.019	0.196	0.979	0.569
175		-1.267	-2.591	-0.212	-0.919	0.331	0.187
176		-4.344	0.557	-0.768	1.289	-1.803	1.484
177		0.275	-0.651	0.111	0.082	0.586	0.495
178		0.320	-0.319	0.485	0.280	0.719	0.598
180		1.000	1.000	1.000	1.000	1.000	1.000
181		-4.011	-9.964	-0.049	-0.873	1.013	0.744

*Not calculable, or estimated as 1,000 or higher because of low expenditure growth.

of values below 0.5. The sectors for which the combined income and composition effect is strongest include 35 and 36 (fabrics), 41 and 46 (lumber
and furniture), 60 (pesticides), 66 (drugs), 76 (shoes), 80 (pottery), 83,
95, 96, 101, 103, 109 and 119 (metal and machinery sectors), 127 (communi-
cation equipment), 134 and 137 (aircraft and boats), 143 and 144 (medical
goods), 151, 153, 155 and 157 (travel), 172 (advertising) and 175 (medical
services). This list includes all the health-related sectors, but not
those related to education: for the latter, income and composition pull
in different directions, so their joint effect is smaller. (The elasticity between scenarios D and E is _negative_ for private education.) It
should also be noted that health care is the only large category for which
the cross-scenario elasticities remain below 0.5 out to the year 2025 (see
table 1-7). Most of the non-health care sector with large combined income
and composition effects take only small shares of total PCE, so they do
not affect the large categories of which they are' part.

Per Capita Spending and Structural Differences

We described, in tables 6-2 through 6-8 above, how the percentage
composition of PCE is projected to change through time in each population
and income scenario. An alternative way to remove the effect of population
scale is to express expenditures in dollars per person. This is done in
tables 6-17 (for the year 1985), 6-18 (for 2000) and 6-19 (for 2025): the
tables are arranged for comparison across scenarios rather than through
time. For scenarios D and F, per capita spending in each sector is compared to that in scenario E, and the percentage difference is shown.

Even in the year 2025, and assuming income growth, there are only two
sectors for which spending per person exceeds $1,000: these are sectors
164 (retail trade) and 175 (medical services). Spending in sector 167
(owner-occupied dwellings) is nearly as large. Another 12 sectors each
absorb between $100 and $1,000: these are 23 (meat products), 39 (clothing), 66 (drugs), 133 (vehicles), 158 (telephone and telegraph), 160 (electricity), 163 (wholesale trade), 168 (real estate services), 170 (personal
and repair services), 173 (auto repair), 174 (amusements) and 176 (private
schools). Ten of these 15 largest sectors are included in the three lar-

Table 6-17
Per Capita Expenditure and
Percentage Difference from Scenario E
by Sector, Scenario and
Income Growth Assumption: 1985

Income Assumption

| | Static | | | | | Growth | | | | |
| | Per Capita Spending | | | Percent Difference | | Per Capita Spending | | | Percent Difference | |
Sector	D	E	F	D	F	D	E	F	D	F
1	0.52	0.52	0.52	-0.00	-0.00	0.20	0.21	0.22	-3.38	2.07
2	6.94	7.18	7.33	-3.38	2.07	6.61	6.68	6.72	-1.06	0.65
3	1.19	1.19	1.19	-0.00	0.00	1.47	1.49	1.50	-1.16	1.24
7	21.03	21.86	22.35	-3.77	2.26	22.05	22.19	22.63	-0.60	1.99
8	2.92	2.92	2.92	0.00	0.00	2.01	2.22	2.50	-9.48	12.58
10	0.14	0.14	0.14	-0.00	-0.00	0.18	0.18	0.18	-0.80	0.94
14	0.00	0.00	0.00	*******	*******	0.00	0.00	0.00	*******	*******
16	0.10	0.10	0.10	0.00	-0.00	0.13	0.14	0.14	-3.07	3.64
21	0.09	0.20	0.33	-55.70	65.95	0.09	0.20	0.33	-55.70	65.95
22	0.19	0.27	0.36	-28.70	32.55	0.19	0.27	0.36	-28.70	32.55
23	100.06	103.56	105.71	-3.38	2.07	113.72	114.35	116.14	-0.55	1.56
24	51.19	52.98	54.08	-3.38	2.07	52.53	52.98	53.02	-0.86	0.06
25	52.02	54.03	55.28	-3.73	2.31	54.55	54.85	55.64	-0.54	1.45
26	17.14	17.79	18.19	-3.64	2.25	17.96	17.99	18.06	-0.17	0.42
27	37.09	38.53	39.41	-3.73	2.30	39.04	39.10	39.56	-0.15	1.18
28	5.11	5.29	5.40	-3.38	2.07	5.26	5.29	5.26	-0.55	-0.51
29	14.48	15.04	15.39	-3.71	2.35	14.01	14.02	14.12	-0.13	0.70
30	55.72	57.95	59.30	-3.83	2.33	58.23	59.01	60.46	-1.33	2.46
31	23.11	23.92	24.41	-3.38	2.07	23.51	23.92	24.34	-1.69	1.79
32	5.79	6.01	6.15	-3.59	2.30	5.81	5.86	5.91	-0.78	0.85
33	27.61	28.67	29.32	-3.71	2.26	28.81	29.06	29.45	-0.87	1.34
34	36.50	37.80	38.60	-3.45	2.12	36.92	37.91	38.89	-2.61	2.59
35	4.71	4.89	4.99	-3.59	2.20	4.74	4.93	4.83	-3.80	-2.01
36	7.55	7.55	7.55	0.00	0.00	9.42	9.92	10.54	-4.96	6.26
37	0.73	0.76	0.77	-3.56	2.37	0.76	0.77	0.79	-0.92	3.43
38	16.09	16.76	17.18	-4.01	2.47	17.23	17.21	17.55	0.17	1.97
39	83.51	87.39	89.79	-4.44	2.74	92.05	91.30	94.64	0.83	3.67
40	12.24	12.84	13.20	-4.61	2.85	13.59	13.51	13.79	0.64	2.12
41	0.92	0.92	0.92	0.00	-0.00	0.95	0.98	1.01	-2.95	3.49
43	1.56	1.63	1.68	-4.60	2.84	1.73	1.72	1.82	0.66	5.73
45	22.87	23.99	24.66	-4.65	2.79	28.46	28.39	29.27	0.23	3.09
46	1.20	1.20	1.20	-0.00	-0.00	1.26	1.34	1.43	-6.22	6.65
48	0.23	0.23	0.23	0.00	0.00	0.25	0.26	0.26	-1.62	1.92
49	11.14	11.55	11.81	-3.57	2.25	11.51	11.66	11.86	-1.26	1.67
51	0.55	0.55	0.55	0.00	0.00	0.73	0.74	0.76	-2.19	2.49
52	10.72	11.10	11.33	-3.38	2.07	10.76	11.10	11.29	-3.06	1.74
53	5.59	5.84	6.00	-4.35	2.68	5.91	6.08	6.49	-2.90	6.69
54	7.59	7.94	8.15	-4.37	2.69	8.55	8.27	8.88	3.38	7.35
55	0.97	0.97	0.97	-0.00	0.00	2.61	2.66	2.69	-1.65	1.07
56	0.47	0.49	0.50	-3.52	2.26	0.50	0.49	0.48	0.94	-2.19
57	1.13	1.18	1.20	-3.72	2.20	1.14	1.19	1.23	-4.46	3.07

Table 6-17 cont'd.

Income Assumption

Sector	Static					Growth				
	Per Capita Spending			Percent Difference		Per Capita Spending			Percent Difference	
	D	E	F	D	F	D	E	F	D	F
58	3.62	3.76	3.85	-3.67	2.28	3.70	3.81	3.95	-2.88	3.71
59	0.28	0.29	0.30	-4.02	2.53	0.29	0.30	0.31	-2.77	4.09
60	0.04	0.04	0.04	0.00	0.01	0.06	0.06	0.07	-4.47	4.97
61	2.34	2.43	2.48	-3.52	2.23	2.75	2.79	2.83	-1.51	1.63
62	0.13	0.13	0.13	0.00	-0.00	0.14	0.15	0.16	-4.10	4.64
66	37.45	38.15	39.65	-1.82	3.95	54.59	59.42	64.18	-8.14	8.01
67	32.03	33.23	33.96	-3.60	2.20	33.01	33.53	34.09	-1.55	1.67
68	0.15	0.15	0.15	0.00	0.00	0.19	0.20	0.20	-1.69	2.00
69	51.71	51.72	51.72	-0.01	0.01	43.32	44.78	45.65	-3.26	1.95
70	14.32	14.32	14.32	0.00	0.00	10.60	10.60	10.60	-0.01	-0.01
72	12.42	12.93	13.25	-3.97	2.44	13.18	13.25	13.77	-0.49	3.92
73	3.64	3.78	3.86	-3.67	2.26	4.02	3.83	3.46	4.98	-9.61
74	2.25	2.33	2.39	-3.64	2.23	2.90	2.90	2.95	-0.05	1.57
76	18.36	19.14	19.61	-4.06	2.50	19.69	19.68	20.04	0.06	1.82
77	5.56	5.83	5.99	-4.55	2.81	6.04	6.11	5.94	-1.15	-2.91
78	2.21	2.21	2.21	0.00	0.00	2.44	2.49	2.58	-2.31	3.35
80	0.74	0.77	0.79	-4.25	2.62	0.77	0.80	0.70	-3.55	-12.18
81	0.02	0.02	0.02	-3.79	2.33	0.02	0.02	0.02	0.05	1.31
82	0.69	0.72	0.74	-3.73	1.99	0.75	0.73	0.71	3.63	-2.46
83	0.07	0.07	0.07	0.00	0.00	0.05	0.06	0.06	-10.44	13.75
87	0.10	0.10	0.10	-3.70	2.18	0.10	0.10	0.10	-3.08	-0.37
90	0.02	0.02	0.02	3.00	-0.00	0.02	0.02	0.02	-7.95	12.50
93	0.00	0.00	0.00	-0.11	0.00	0.02	0.02	0.02	-2.95	1.90
94	0.00	0.00	0.22	*******	*******	0.00	0.00	0.00	*******	*******
95	0.19	0.19	0.19	-0.00	-0.00	0.26	0.27	0.29	-6.05	6.76
96	0.23	0.23	0.23	-0.00	0.00	0.17	0.19	0.21	-8.08	10.59
97	2.29	2.38	2.43	-3.63	2.23	2.41	2.40	2.27	0.43	-5.46
98	4.23	4.23	4.23	0.00	-0.00	4.59	4.73	4.90	-2.78	3.69
99	0.21	0.21	0.21	-0.00	-0.00	0.14	0.15	0.16	-5.52	7.32
101	0.93	0.93	0.93	-0.00	-0.00	1.31	1.37	1.44	-4.33	5.43
102	1.18	1.18	1.18	0.00	0.00	1.07	1.17	1.28	-8.85	9.41
103	3.05	3.18	3.26	-4.03	2.48	3.25	3.27	3.31	-0.62	1.27
106	0.17	0.17	0.17	0.00	0.00	0.23	0.25	0.26	-6.04	7.15
108	0.62	0.62	0.62	-0.00	0.00	0.75	0.78	0.82	-3.81	5.25
109	0.17	0.17	0.17	0.00	-0.00	0.15	0.16	0.17	-6.84	7.32
115	0.83	0.83	0.83	-0.00	0.00	1.18	1.25	1.33	-5.64	6.67
116	4.66	4.85	4.96	-3.93	2.42	4.89	4.96	4.97	-1.29	0.24
117	0.02	0.02	0.02	-0.00	-0.00	0.04	0.04	0.04	-5.24	5.71
119	0.07	0.07	0.07	0.01	0.00	0.08	0.08	0.10	-10.20	12.07
120	0.13	0.13	0.13	0.00	0.00	0.14	0.16	0.17	-9.43	11.65
122	0.00	0.01	0.01	-4.02	2.50	0.01	0.01	0.00	5.24	-6.69
123	29.05	30.25	30.99	-3.97	2.45	30.54	31.00	31.93	-1.50	2.99
124	4.19	4.37	4.49	-4.31	2.61	4.53	4.54	4.58	-0.27	0.90
125	26.85	27.99	28.69	-4.05	2.50	30.83	31.11	32.26	-0.88	3.69
126	2.55	2.66	2.73	-4.14	2.55	2.72	2.75	2.83	-0.85	2.93
127	0.80	0.80	0.80	0.00	0.00	1.14	1.23	1.33	-6.88	8.14

Table 6-17 cont'd.

Income Assumption

Sector	Static Per Capita Spending D	E	F	Static Percent Difference D	F	Growth Per Capita Spending D	E	F	Growth Percent Difference D	F
128	1.32	1.38	1.41	-4.41	2.72	1.45	1.44	1.53	0.68	6.61
129	2.34	2.34	2.34	0.00	-0.00	2.64	2.76	2.90	-4.42	4.98
130	0.83	0.86	0.88	-3.84	2.36	0.87	0.87	0.91	-0.82	3.52
131	0.43	0.45	0.46	-4.01	2.47	0.48	0.46	0.43	4.26	-6.17
133	150.76	158.90	163.92	-5.12	3.16	161.79	159.91	171.61	1.18	7.32
134	1.19	1.19	1.19	-0.00	0.00	1.97	2.24	2.52	-11.97	12.73
137	2.38	2.38	2.38	0.00	0.00	3.54	3.73	3.96	-5.05	5.98
139	4.75	4.98	5.11	-4.47	2.76	5.39	5.21	5.83	3.57	12.08
140	9.75	10.45	10.76	-6.70	3.02	10.99	11.20	12.16	-1.81	8.65
142	0.12	0.13	0.13	-3.98	2.45	0.13	0.13	0.15	-1.23	10.21
143	2.16	2.20	2.29	-1.87	4.02	3.15	3.43	3.71	-8.18	8.08
144	1.08	1.10	1.15	-1.87	4.02	1.57	1.71	1.85	-8.18	8.08
145	5.18	5.41	5.55	-4.27	2.63	5.58	5.61	5.91	-0.57	5.26
146	3.11	3.26	3.36	-4.74	2.93	3.12	3.12	3.20	0.08	2.41
147	11.03	11.68	12.09	-5.59	3.47	10.46	10.19	10.69	2.66	4.90
148	17.47	18.22	18.69	-4.14	2.58	21.77	21.43	21.10	1.59	-1.55
149	1.39	1.39	1.39	0.00	0.00	1.68	1.75	1.81	-3.49	3.70
150	4.25	4.25	4.25	-0.00	-0.00	3.95	4.18	4.45	-5.35	6.45
151	2.15	2.23	2.29	-3.95	2.43	2.21	2.29	2.08	-3.47	-9.13
152	13.88	14.43	14.77	-3.82	2.36	14.64	14.71	15.51	-0.50	5.41
153	4.00	4.19	4.30	-4.40	2.71	4.24	4.36	4.38	-2.78	0.44
154	1.32	1.40	1.45	-5.75	3.57	1.17	1.20	1.18	-2.91	-2.03
155	19.58	20.67	21.35	-5.31	3.29	22.65	22.36	23.28	1.29	4.07
157	0.09	0.10	0.10	-4.05	2.49	0.10	0.10	0.10	-5.27	2.24
158	59.55	61.96	63.45	-3.90	2.40	68.67	70.06	72.29	-1.98	3.19
160	56.85	56.85	56.85	-0.00	0.00	101.12	101.15	101.22	-0.03	0.07
161	28.88	28.88	28.88	-0.00	-0.00	19.73	19.73	19.73	-0.01	-0.01
162	14.04	14.61	14.96	-3.87	2.38	14.80	14.91	15.26	-0.72	2.33
163	206.17	212.78	217.52	-3.11	2.23	225.33	229.89	238.34	-1.99	3.67
164	615.89	635.63	649.79	-3.11	2.23	673.11	686.75	711.98	-1.99	3.67
165	102.64	107.50	110.50	-4.52	2.79	114.92	112.66	117.85	2.01	4.61
166	80.30	83.74	85.86	-4.11	2.53	85.83	86.30	89.68	-0.54	3.93
167	338.53	338.53	338.53	-0.00	-0.00	440.58	448.29	457.42	-1.72	2.04
168	177.91	184.70	188.88	-3.68	2.26	195.80	204.87	210.83	-4.42	2.91
169	19.16	20.27	20.96	-5.49	3.38	22.92	22.08	23.58	3.83	6.79
170	78.30	81.90	84.12	-4.39	2.71	85.04	85.40	88.82	-0.43	4.00
171	28.22	29.53	30.34	-4.43	2.74	30.62	30.85	32.14	-0.77	4.15
172	1.12	1.12	1.12	0.00	-0.00	1.42	1.49	1.56	-4.74	5.04
173	56.88	56.88	56.88	0.00	0.00	62.00	63.51	65.22	-2.39	2.68
174	36.32	38.26	39.46	-5.06	3.13	42.01	40.99	42.53	2.49	3.77
175	230.33	232.96	241.33	-1.13	3.59	335.70	362.87	390.63	-7.49	7.65
176	42.43	40.05	39.64	5.94	-1.02	54.33	65.27	65.87	-16.77	0.91
177	8.96	9.30	9.51	-3.65	2.25	9.18	9.41	9.74	-2.46	3.45
178	0.05	0.05	0.05	-0.00	-0.00	0.05	0.05	0.06	-2.31	2.74
180	3.06	3.06	3.06	0.00	0.00	4.37	4.37	4.37	-0.00	0.00
181	6.82	6.82	6.82	-0.00	0.00	5.26	6.25	7.82	-15.81	25.23

Table 6-18
Per Capita Expenditure and
Percentage Difference from Scenario E
by Sector, Scenario and
Income Growth Assumption: 2000

	Income Assumption									
	Static					Growth				
	Per Capita Spending			Percent Difference		Per Capita Spending			Percent Difference	
Sector	D	E	F	D	F	D	E	F	D	F
1	0.52	0.52	0.52	-0.00	-0.00	0.17	0.19	0.20	-7.53	5.48
2	6.79	7.34	7.75	-7.53	5.48	6.25	6.41	6.67	-2.47	4.08
3	1.19	1.19	1.19	0.00	-0.00	2.00	2.04	2.08	-1.86	1.82
7	21.64	22.36	23.40	-3.19	4.67	25.95	27.21	28.41	-4.64	4.41
8	2.92	2.92	2.92	-0.00	-0.00	2.32	2.75	3.19	-15.68	15.92
10	0.14	0.14	0.14	0.00	-0.00	0.24	0.25	0.25	-1.27	1.26
14	0.00	0.00	0.00	*******	*******	0.00	0.00	0.00	*******	*******
16	0.10	0.10	0.10	-0.00	-0.00	0.18	0.19	0.20	-4.60	4.57
21	0.00	0.00	0.00	*******	*******	0.00	0.00	0.00	*******	*******
22	0.00	0.00	0.00	*******	*******	0.00	0.00	0.01	*******	*******
23	97.77	105.73	111.53	-7.53	5.48	119.95	122.29	127.47	-1.92	4.23
24	49.48	53.51	56.45	-7.53	5.48	52.53	53.51	55.51	-1.84	3.74
25	53.09	54.93	57.52	-3.35	4.70	63.79	66.32	69.04	-3.82	4.09
26	17.30	18.06	18.94	-4.21	4.87	20.51	21.07	21.94	-2.66	4.12
27	37.87	39.16	41.00	-3.30	4.71	46.05	47.71	49.58	-3.48	3.93
28	4.90	5.30	5.59	-7.53	5.48	5.21	5.30	5.49	-1.69	3.71
29	14.75	15.24	15.95	-3.19	4.69	14.98	15.49	16.11	-3.30	3.98
30	58.56	59.93	62.64	-2.30	4.52	71.81	76.04	79.31	-5.57	4.29
31	22.58	24.42	25.75	-7.53	5.48	23.77	24.42	25.37	-2.66	3.92
32	5.77	6.05	6.35	-4.62	4.94	6.15	6.38	6.63	-3.56	4.07
33	28.32	29.42	30.83	-3.75	4.78	33.38	35.00	36.47	-4.62	4.21
34	37.11	39.80	41.93	-6.76	5.34	39.20	41.25	43.14	-4.97	4.57
35	4.54	4.78	5.02	-5.13	5.02	4.83	5.29	5.45	-8.65	3.06
36	7.55	7.55	7.55	0.00	0.00	13.71	14.73	15.75	-6.89	6.93
37	0.75	0.79	0.83	-4.51	4.92	0.86	0.90	0.94	-4.14	4.44
38	16.97	16.98	17.68	-0.05	4.10	22.45	23.47	24.42	-4.34	4.05
39	94.08	89.38	92.24	5.26	3.20	142.69	152.90	159.59	-6.68	4.38
40	13.98	12.96	13.32	7.84	2.79	22.72	24.48	25.66	-7.19	4.80
41	0.92	0.92	0.92	0.00	-0.00	1.17	1.23	1.29	-5.13	5.10
43	1.79	1.66	1.71	7.44	2.79	2.82	3.02	3.19	-6.50	5.70
45	25.80	23.99	24.65	7.53	2.74	37.28	40.34	42.26	-7.58	4.75
46	1.20	1.20	1.20	-0.00	0.00	1.85	2.04	2.22	-9.04	8.79
48	0.23	0.23	0.23	-0.00	0.00	0.29	0.30	0.31	-2.99	2.97
49	11.24	11.84	12.44	-5.08	5.02	12.75	13.25	13.80	-3.77	4.16
51	0.55	0.55	0.55	0.00	0.00	1.01	1.05	1.08	-3.42	3.37
52	10.63	11.50	12.13	-7.53	5.48	10.88	11.50	11.99	-5.35	4.27
53	6.54	6.28	6.49	4.14	3.38	9.07	10.25	10.82	-11.53	5.60
54	8.06	7.70	7.96	4.68	3.28	12.63	12.79	13.30	-1.25	4.00
55	0.97	0.97	0.97	-0.00	0.00	4.28	4.39	4.48	-2.47	1.99
56	0.45	0.48	0.50	-5.32	5.09	0.54	0.54	0.55	-0.08	2.29
57	1.21	1.26	1.32	-4.23	4.87	1.33	1.47	1.54	-8.95	4.95

Table 6-18 cont'd.

Income Assumption

Sector	Static					Growth				
	Per Capita Spending			Percent Difference		Per Capita Spending			Percent Difference	
	D	E	F	D	F	D	E	F	D	F
58	3.80	3.97	4.16	-4.26	4.88	4.34	4.63	4.85	-6.10	4.80
59	0.31	0.31	0.32	0.40	4.05	0.39	0.44	0.46	-9.57	5.37
60	0.04	0.04	0.04	0.01	-0.00	0.10	0.10	0.11	-6.02	5.92
61	2.36	2.49	2.62	-5.48	5.10	3.45	3.63	3.78	-4.86	4.14
62	0.13	0.13	0.13	0.00	0.00	0.22	0.23	0.25	-5.71	5.63
66	37.36	38.78	40.97	-3.67	5.64	90.45	100.73	112.67	-10.21	11.85
67	32.42	34.11	35.82	-4.96	5.00	36.80	38.50	40.13	-4.40	4.25
68	0.15	0.15	0.15	0.00	-0.00	0.25	0.26	0.27	-2.76	2.74
69	51.71	51.72	51.72	-0.01	0.01	37.20	38.66	39.83	-3.76	3.03
70	14.32	14.32	14.32	0.00	-0.00	8.75	8.74	8.73	0.10	-0.09
72	13.17	13.28	13.83	-0.83	4.21	16.63	17.43	18.18	-4.58	4.32
73	3.36	3.52	3.69	-4.50	4.84	4.30	3.99	4.02	7.75	0.71
74	2.28	2.39	2.51	-4.56	4.93	4.17	4.28	4.45	-2.54	3.95
76	19.55	19.43	20.20	0.63	3.99	26.23	27.69	28.83	-5.27	4.15
77	5.97	5.57	5.73	7.13	2.89	9.10	10.18	10.53	-10.61	3.46
78	2.21	2.21	2.21	0.00	-0.00	3.33	3.45	3.57	-3.36	3.47
80	0.66	0.64	0.66	3.37	3.37	0.83	0.95	0.95	-12.44	0.00
81	0.02	0.02	0.02	-3.01	4.61	0.02	0.03	0.03	-4.20	3.36
82	0.63	0.67	0.70	-5.03	4.99	0.79	0.75	0.76	5.62	1.96
83	0.07	0.07	0.07	-0.00	-0.00	0.07	0.08	0.09	-15.13	15.35
87	0.10	0.10	0.11	-4.41	4.89	0.11	0.12	0.12	-7.49	4.11
90	0.02	0.02	0.02	-0.01	0.00	0.02	0.02	0.02	-12.09	12.77
93	0.00	0.00	0.00	-0.10	-0.02	0.04	0.04	0.04	-3.76	3.03
94	0.00	0.00	0.22	*******	*******	0.00	0.00	0.00	*******	*******
95	0.19	0.19	0.19	0.00	0.00	0.43	0.46	0.50	-7.61	7.49
96	0.23	0.23	0.23	0.00	0.00	0.27	0.30	0.33	-10.56	10.69
97	2.06	2.18	2.29	-5.28	5.04	2.49	2.39	2.42	4.11	1.14
98	4.23	4.23	4.23	-0.00	-0.00	6.45	6.72	7.00	-4.06	4.12
99	0.21	0.21	0.21	0.00	-0.00	0.12	0.14	0.15	-12.51	12.70
101	0.93	0.93	0.93	-0.00	-0.00	2.15	2.28	2.41	-5.50	5.52
102	1.18	1.18	1.18	0.00	-0.00	1.64	1.87	2.10	-12.28	11.99
103	3.29	3.28	3.41	0.34	4.02	4.31	4.56	4.79	-5.54	4.92
106	0.17	0.17	0.17	0.00	-0.00	0.34	0.37	0.41	-8.43	8.38
108	0.62	0.62	0.62	0.00	-0.00	1.12	1.18	1.24	-5.15	5.27
109	0.17	0.17	0.17	-0.00	0.00	0.22	0.24	0.27	-9.94	9.72
115	0.83	0.83	0.83	-0.00	0.00	1.97	2.12	2.27	-7.14	7.09
116	4.80	4.87	5.08	-1.52	4.32	5.80	6.16	6.37	-5.77	3.47
117	0.02	0.02	0.02	-0.00	-0.01	0.06	0.07	0.07	-6.59	6.47
119	0.07	0.07	0.07	-0.00	-0.00	0.13	0.15	0.17	-12.26	12.18
120	0.13	0.13	0.13	-0.00	-0.00	0.25	0.28	0.31	-11.11	11.13
122	0.00	0.00	0.00	1.37	3.75	0.01	0.01	0.01	6.07	2.62
123	31.18	31.35	32.66	-0.53	4.18	39.54	42.22	44.15	-6.35	4.58
124	4.59	4.41	4.56	4.09	3.44	6.82	7.35	7.70	-7.16	4.78
125	28.99	28.89	30.06	0.34	4.03	45.07	47.73	49.80	-5.57	4.33
126	2.69	2.64	2.75	1.66	3.80	3.71	3.91	4.06	-5.11	3.96
127	0.80	0.80	0.80	-0.00	-0.00	1.95	2.13	2.31	-8.60	8.55

238

Table 6-18 cont'd.

Income Assumption

Sector	Static					Growth				
	Per Capita Spending			Percent Difference		Per Capita Spending			Percent Difference	
	D	E	F	D	F	D	E	F	D	F
128	1.45	1.38	1.42	5.13	3.19	2.19	2.31	2.43	-5.29	4.96
129	2.34	2.34	2.34	-0.00	-0.00	6.01	6.27	6.53	-4.19	4.13
130	0.86	0.88	0.92	-2.46	4.50	1.03	1.08	1.13	-4.43	4.31
131	0.41	0.41	0.43	0.69	3.89	0.59	0.56	0.57	4.74	2.68
133	180.76	158.90	161.57	13.76	1.68	206.42	225.77	237.58	-8.57	5.23
134	1.19	1.19	1.19	0.00	0.00	3.44	4.04	4.62	-14.69	14.34
137	2.38	2.38	2.38	-0.00	-0.00	7.20	7.70	8.21	-6.54	6.50
139	5.60	5.29	5.45	5.92	3.02	8.91	8.97	9.46	-0.68	5.49
140	8.93	8.90	8.86	0.29	-0.53	13.78	13.29	15.30	3.72	15.18
142	0.14	0.14	0.15	-0.80	4.18	0.17	0.18	0.19	-5.63	5.57
143	2.13	2.20	2.31	-3.08	5.23	5.16	5.71	6.36	-9.66	11.41
144	1.07	1.10	1.16	-3.08	5.23	2.58	2.86	3.18	-9.66	11.41
145	5.83	5.66	5.86	3.03	3.56	8.26	8.82	9.25	-6.31	4.93
146	3.60	3.29	3.38	9.39	2.54	4.78	5.27	5.49	-9.32	4.18
147	14.42	11.75	11.79	22.78	0.34	18.89	20.68	21.92	-8.66	6.00
148	18.28	18.00	18.69	1.53	3.84	34.31	35.23	36.34	-2.61	3.15
149	1.39	1.39	1.39	0.00	0.00	2.41	2.55	2.68	-5.27	5.10
150	4.25	4.25	4.25	0.00	-0.00	5.19	5.66	6.14	-8.45	8.42
151	2.01	2.03	2.11	-0.95	4.26	2.43	2.69	2.73	-9.68	1.56
152	14.77	15.10	15.79	-2.20	4.59	18.98	20.15	21.02	-5.78	4.31
153	4.48	4.28	4.42	4.58	3.31	6.15	7.13	7.41	-13.71	3.89
154	1.67	1.31	1.31	27.00	-0.35	2.05	2.44	2.60	-16.02	6.38
155	24.38	20.91	21.17	16.64	1.25	44.79	50.12	52.62	-10.62	4.99
157	0.09	0.09	0.10	-0.68	4.16	0.11	0.12	0.13	-14.17	2.16
158	63.40	64.33	67.14	-1.46	4.36	97.69	104.79	109.62	-6.77	4.61
160	56.85	56.85	56.85	-0.00	0.00	127.77	127.73	127.79	0.03	0.04
161	28.88	28.88	28.88	0.00	0.00	12.61	12.59	12.58	0.10	-0.09
162	14.90	15.15	15.81	-1.64	4.39	18.69	19.64	20.59	-4.85	4.83
163	227.57	224.09	230.96	1.55	3.07	318.95	341.25	361.03	-6.54	5.80
164	679.82	669.42	689.94	1.55	3.07	952.78	1019.42	1078.50	-6.54	5.80
165	117.64	109.81	112.95	7.13	2.86	187.83	198.35	208.59	-5.30	5.17
166	87.79	86.86	90.25	1.08	3.91	117.61	125.74	131.60	-6.47	4.66
167	338.53	338.53	338.53	-0.00	-0.00	635.18	652.04	668.79	-2.59	2.57
168	185.15	192.65	201.98	-3.90	4.84	256.12	283.07	294.27	-9.52	3.96
169	24.05	20.29	20.48	18.52	0.93	46.44	51.01	53.18	-8.96	4.26
170	89.60	85.75	88.62	4.50	3.35	131.58	143.64	150.42	-8.39	4.72
171	31.76	30.01	30.96	5.84	3.15	48.62	53.33	55.83	-8.84	4.68
172	1.12	1.12	1.12	-0.00	-0.00	2.24	2.39	2.55	-6.53	6.38
173	56.88	56.88	56.88	0.00	-0.00	79.43	83.11	86.64	-4.42	4.25
174	43.62	38.49	39.19	13.33	1.83	78.52	84.79	88.51	-7.39	4.38
175	226.86	232.52	243.27	-2.44	4.62	549.27	603.93	669.07	-9.05	10.79
176	40.85	37.68	35.81	8.43	-4.95	87.44	100.42	98.89	-12.93	-1.53
177	9.31	9.73	10.21	-4.30	4.89	10.59	11.35	11.92	-6.73	5.02
178	0.05	0.05	0.05	-0.00	-0.00	0.06	0.07	0.07	-3.95	3.92
180	3.06	3.06	3.06	0.00	0.00	6.07	6.07	6.07	0.00	0.00
181	6.82	6.82	6.82	0.00	0.00	18.37	19.94	22.04	-7.89	10.51

Table 6-19
Per Capita Expenditure and
Percentage Difference from Scenario E
by Sector, Scenario and
Income Growth Assumption: 2025

Income Assumption

	Static					Growth				
	Per Capita Spending			Percent Difference		Per Capita Spending			Percent Difference	
Sector	D	E	F	D	F	D	E	F	D	F
1	0.52	0.52	0.52	0.00	0.00	0.14	0.16	0.19	-17.33	14.68
2	6.30	7.62	8.73	-17.33	14.68	5.89	6.13	6.44	-3.93	5.08
3	1.19	1.19	1.19	-0.00	-0.00	2.91	2.98	3.05	-2.22	2.49
7	21.03	23.08	25.25	-8.89	9.38	31.99	33.38	35.36	-4.18	5.94
8	2.92	2.92	2.92	0.00	0.00	1.89	2.58	3.30	-26.76	28.02
10	0.14	0.14	0.14	0.00	0.00	0.36	0.36	0.37	-1.50	1.64
14	0.00	0.00	0.00	*******	*******	0.00	0.00	0.00	*******	*******
16	0.10	0.10	0.10	0.00	-0.00	0.25	0.26	0.28	-5.41	5.89
21	0.00	0.00	0.00	*******	*******	0.00	0.00	0.00	*******	*******
22	0.00	0.00	0.00	*******	*******	0.00	0.00	0.00	*******	*******
23	90.16	109.06	125.08	-17.33	14.68	128.10	133.24	139.75	-3.86	4.89
24	44.71	54.08	62.02	-17.33	14.68	52.80	54.08	56.39	-2.38	4.27
25	49.98	54.94	60.12	-9.03	9.42	76.80	78.68	82.60	-2.39	4.98
26	16.17	18.07	19.96	-10.52	10.42	24.24	24.64	25.69	-1.65	4.26
27	35.34	38.76	42.44	-8.82	9.50	56.10	57.13	59.79	-1.80	4.66
28	4.55	5.50	6.31	-17.33	14.68	5.33	5.50	5.80	-3.11	5.44
29	13.79	15.08	16.49	-8.57	9.37	15.56	15.79	16.54	-1.45	4.76
30	54.95	58.75	63.48	-6.46	8.05	90.52	94.09	99.06	-3.79	5.29
31	19.75	23.89	27.40	-17.33	14.68	23.29	23.89	24.55	-2.49	2.78
32	5.52	6.26	6.95	-11.80	11.04	6.63	6.85	7.24	-3.22	5.71
33	27.18	30.19	33.23	-9.98	10.08	40.35	41.99	44.50	-3.92	5.96
34	33.44	39.72	45.15	-15.81	13.67	40.44	42.60	44.44	-5.06	4.33
35	3.98	4.54	5.05	-12.39	11.19	5.14	5.47	5.86	-6.03	7.16
36	7.55	7.55	7.55	0.00	-0.00	17.15	18.66	20.27	-8.09	8.59
37	0.72	0.81	0.90	-11.40	10.82	1.00	1.05	1.10	-4.53	5.02
38	16.04	16.36	17.22	-1.99	5.24	29.81	30.08	31.48	-0.88	4.66
39	92.83	84.75	84.51	9.54	-0.28	224.66	228.56	241.59	-1.71	5.70
40	14.10	12.25	11.95	15.05	-2.46	38.68	38.90	41.18	-0.56	5.84
41	0.92	0.92	0.92	0.00	0.00	1.46	1.57	1.69	-7.03	7.65
43	1.87	1.69	1.67	10.29	-1.33	4.37	4.49	4.74	-2.58	5.48
45	27.74	23.99	23.17	15.62	-3.40	55.48	56.46	59.50	-1.72	5.40
46	1.20	1.20	1.20	-0.00	-0.00	2.49	2.81	3.16	-11.18	12.39
48	0.23	0.23	0.23	-0.00	-0.00	0.35	0.37	0.39	-4.27	4.65
49	10.21	11.67	13.02	-12.48	11.55	14.14	14.48	15.07	-2.33	4.05
51	0.55	0.55	0.55	0.00	-0.00	1.48	1.54	1.61	-4.08	4.50
52	9.76	11.81	13.54	-17.33	14.68	11.04	11.81	12.44	-6.46	5.36
53	6.67	6.27	6.33	6.41	1.03	14.00	15.27	16.34	-8.29	7.02
54	7.68	7.11	7.11	8.11	0.03	18.62	18.04	18.51	3.24	2.60
55	0.97	0.97	0.97	-0.00	0.00	3.92	3.99	4.03	-1.71	0.87
56	0.39	0.45	0.51	-12.87	12.28	0.60	0.58	0.59	3.59	2.43
57	1.16	1.30	1.43	-10.92	10.64	1.59	1.73	1.85	-7.79	7.03

Table 6-19 cont'd.

Income Assumption

Sector	Static Per Capita Spending			Static Percent Difference		Growth Per Capita Spending			Growth Percent Difference	
	D	E	F	D	F	D	E	F	D	F
58	3.64	4.11	4.56	-11.36	11.01	5.16	5.44	5.75	-5.12	5.75
59	0.31	0.31	0.33	-1.85	5.67	0.56	0.60	0.64	-6.16	6.74
60	0.04	0.04	0.04	0.00	0.00	0.15	0.16	0.17	-6.55	7.27
61	2.20	2.54	2.85	-13.38	12.17	4.90	5.18	5.46	-5.33	5.45
62	0.13	0.13	0.13	-0.00	0.00	0.34	0.36	0.38	-6.37	7.04
66	39.36	42.60	46.51	-7.59	9.19	171.27	198.61	225.11	-13.77	13.34
67	29.73	33.84	37.72	-12.17	11.46	41.59	43.04	45.03	-3.36	4.63
68	0.15	0.15	0.15	0.00	-0.00	0.34	0.36	0.37	-3.42	3.73
69	51.71	51.72	51.72	-0.01	0.01	34.74	36.05	36.72	-3.64	1.85
70	14.32	14.32	14.32	-0.00	0.00	2.82	2.76	2.78	2.19	0.62
72	12.54	13.05	13.78	-3.90	5.66	20.59	21.19	22.21	-2.86	4.77
73	2.85	3.22	3.56	-11.53	10.61	4.66	4.01	4.02	16.38	0.38
74	2.07	2.34	2.60	-11.44	10.93	7.04	7.10	7.36	-0.92	3.53
76	18.66	18.75	19.60	-0.44	4.53	35.96	36.71	38.63	-2.04	5.25
77	5.81	5.08	4.95	14.38	-2.59	15.05	15.54	16.90	-3.17	8.74
78	2.21	2.21	2.21	0.00	-0.00	4.50	4.68	4.87	-3.87	3.90
80	0.62	0.58	0.57	6.32	-2.06	1.07	1.15	1.30	-6.42	13.93
81	0.02	0.02	0.02	-8.41	8.38	0.03	0.03	0.03	-0.25	3.11
82	0.57	0.66	0.73	-12.37	11.28	0.87	0.80	0.83	8.19	2.91
83	0.07	0.07	0.07	0.00	-0.00	0.07	0.09	0.11	-21.54	22.62
87	0.09	0.10	0.11	-11.12	10.59	0.12	0.13	0.14	-5.56	6.48
90	0.02	0.02	0.02	0.00	0.01	0.02	0.02	0.02	-20.01	19.24
93	0.00	0.00	0.00	0.01	0.05	0.02	0.02	0.03	-3.63	1.85
94	0.00	0.00	0.22	*******	*******	0.00	0.00	0.00	*******	*******
95	0.19	0.19	0.19	-0.00	0.00	0.67	0.73	0.79	-8.05	8.92
96	0.23	0.23	0.23	-0.00	-0.00	0.37	0.42	0.48	-12.48	13.13
97	1.78	2.02	2.25	-12.23	11.07	2.71	2.45	2.50	10.81	2.12
98	4.23	4.23	4.23	-0.00	-0.00	8.98	9.42	9.89	-4.72	4.93
99	0.21	0.21	0.21	0.00	-0.00	0.04	0.07	0.10	-38.54	40.36
101	0.93	0.93	0.93	-0.00	0.00	3.40	3.61	3.84	-5.83	6.23
102	1.18	1.18	1.18	0.00	-0.00	2.43	2.83	3.29	-14.29	16.04
103	3.17	3.20	3.34	-0.93	4.37	5.79	5.96	6.25	-2.81	4.85
106	0.17	0.17	0.17	0.00	-0.00	0.49	0.54	0.60	-9.82	10.70
108	0.62	0.62	0.62	-0.00	-0.00	1.61	1.71	1.81	-5.77	5.95
109	0.17	0.17	0.17	0.00	0.00	0.31	0.35	0.40	-11.79	13.22
115	0.83	0.83	0.83	-0.00	0.00	3.11	3.37	3.65	-7.63	8.31
116	4.42	4.68	4.97	-5.40	6.25	6.80	6.99	7.39	-2.63	5.73
117	0.02	0.02	0.02	0.00	0.00	0.10	0.11	0.12	-6.89	7.69
119	0.07	0.07	0.07	0.00	0.00	0.20	0.23	0.26	-13.45	14.65
120	0.13	0.13	0.13	0.00	0.00	0.37	0.43	0.48	-12.06	12.96
122	0.00	0.00	0.00	2.90	0.33	0.01	0.01	0.01	12.28	0.89
123	29.38	30.22	31.87	-2.78	5.48	51.31	53.16	55.79	-3.48	4.94
124	4.48	4.17	4.22	7.59	1.38	10.86	11.07	11.71	-1.93	5.78
125	27.52	27.84	29.11	-1.14	4.57	69.45	71.29	74.64	-2.57	4.71
126	2.50	2.43	2.50	2.64	2.67	5.17	5.18	5.40	-0.28	4.14
127	0.80	0.80	0.80	0.00	0.00	3.10	3.42	3.76	-9.27	10.11

Table 6-19 cont'd.

Income Assumption

Sector	Static					Growth				
	Per Capita Spending			Percent Difference		Per Capita Spending			Percent Difference	
	D	E	F	D	F	D	E	F	D	F
128	1.39	1.29	1.28	7.90	-0.25	3.15	3.17	3.28	-0.78	3.49
129	2.34	2.34	2.34	0.00	-0.00	114.45	114.90	115.39	-0.39	0.43
130	0.81	0.87	0.94	-7.17	7.50	1.19	1.24	1.30	-3.75	4.73
131	0.37	0.36	0.37	1.23	1.57	0.72	0.65	0.65	10.96	1.14
133	204.86	158.90	139.20	28.92	-12.40	309.75	320.17	338.85	-3.25	5.84
134	1.19	1.19	1.19	-0.00	0.00	5.47	6.51	7.69	-16.04	18.01
137	2.38	2.38	2.38	0.00	-0.00	13.00	14.00	15.10	-7.17	7.81
139	5.58	4.97	4.81	12.20	-3.22	13.37	13.20	13.30	1.30	0.78
140	9.00	8.49	8.22	6.01	-3.07	20.35	17.85	22.67	14.01	27.02
142	0.13	0.13	0.14	-3.21	4.70	0.20	0.21	0.22	-5.04	2.18
143	2.23	2.39	2.58	-6.51	8.20	9.71	11.13	12.50	-12.76	12.31
144	1.12	1.19	1.29	-6.51	8.20	4.85	5.56	6.25	-12.76	12.31
145	5.61	5.34	5.42	5.06	1.46	12.10	12.32	12.87	-1.73	4.53
146	3.57	2.99	2.86	19.21	-4.38	6.77	6.92	7.39	-2.19	6.75
147	15.58	9.71	7.55	60.54	-22.19	25.97	26.34	27.76	-1.39	5.39
148	16.51	16.23	16.79	1.71	3.43	60.59	58.12	60.27	4.24	3.69
149	1.39	1.39	1.39	0.00	0.00	3.29	3.51	3.76	-6.33	6.98
150	4.25	4.25	4.25	-0.00	-0.00	6.54	7.35	8.22	-11.04	11.96
151	2.30	2.44	2.61	-5.73	7.09	3.62	3.83	4.39	-5.52	14.59
152	14.70	15.82	17.36	-7.06	9.77	26.68	27.93	29.72	-4.46	6.44
153	4.46	4.15	4.18	7.52	0.65	9.58	10.55	11.65	-9.20	10.39
154	1.76	0.98	0.68	79.48	-30.54	2.86	3.02	3.29	-5.42	8.89
155	27.15	20.44	18.01	32.84	-11.88	85.74	89.32	96.31	-4.01	7.82
157	0.08	0.09	0.09	-4.85	5.86	0.12	0.13	0.14	-10.83	8.51
158	60.98	64.22	68.78	-5.05	7.11	154.62	162.13	171.45	-4.63	5.75
160	56.85	56.85	56.85	-0.00	0.00	161.94	159.38	161.80	1.60	1.52
161	28.88	28.88	28.88	-0.00	-0.00	2.91	2.85	2.87	2.19	0.62
162	14.52	15.38	16.52	-5.59	7.39	24.27	25.12	26.47	-3.38	5.35
163	242.22	236.79	242.35	2.29	2.35	549.98	576.23	615.26	-4.56	6.77
164	723.57	707.36	723.98	2.29	2.35	1642.95	1721.36	1837.96	-4.56	6.77
165	118.07	101.88	98.08	15.89	-3.73	310.58	316.91	330.93	-2.00	4.43
166	85.82	85.74	89.35	0.09	4.20	164.06	170.38	180.04	-3.71	5.67
167	338.53	338.53	338.53	0.00	-0.00	962.37	991.83	1023.94	-2.97	3.24
168	170.09	187.90	207.48	-9.48	10.42	378.87	411.60	438.44	-7.95	6.52
169	26.75	19.63	16.98	36.27	-13.49	88.97	90.89	97.31	-2.11	7.07
170	91.05	85.20	86.35	6.88	1.36	209.37	217.34	232.07	-3.67	6.77
171	30.89	27.49	27.24	12.36	-0.89	80.13	82.91	87.70	-3.35	5.78
172	1.12	1.12	1.12	-0.00	-0.00	3.55	3.82	4.13	-7.21	8.09
173	56.88	56.88	56.88	0.00	0.00	107.00	113.35	120.60	-5.60	6.40
174	45.19	35.25	31.96	28.19	-9.32	142.91	143.48	152.22	-0.40	6.09
175	237.37	251.61	270.97	-5.66	7.69	1032.82	1173.18	1311.38	-11.96	11.78
176	40.96	34.74	31.96	17.88	-8.02	103.21	175.96	164.68	-41.35	-6.41
177	9.22	10.39	11.54	-11.27	11.02	12.83	13.88	14.87	-7.57	7.16
178	0.05	0.05	0.05	-0.00	-0.00	0.08	0.08	0.09	-5.20	5.66
180	3.06	3.06	3.06	-0.00	-0.00	9.06	9.06	9.06	0.00	-0.00
181	6.82	6.82	6.82	0.00	-0.00	37.81	37.71	39.06	0.25	3.57

242

gest of the categories analyzed in chapter 1, health care, trade, and "other services." Spending per person falls between $10 and $100 in another 40 sectors, and is below $10 in all the remaining 78 sectors.

When per capita expenditure by sector is compared across the three population scenarios, the chief result is that the differences are quite small. Spending per person is nearly always least in scenario D and greatest in F, reflecting the differences in total PCE per capita, which are as follows (percentage differences are shown in parentheses):

Scenario

Year	D	E	F
1975	3258 (0)	3258 (0)	3258
	(18.39)	(22.53)	(27.52)
1985	3857 (-3.38)	3992 (3.91)	4148
	(42.55)	(44.99)	(46.46)
2000	5498 (-5.01)	5788 (4.96)	6075
	(55.93)	(56.74)	(58.41)
2025	8573 (-5.50)	9072 (6.08)	9624

This pattern is reversed--spending per person is highest in D and lowest in F--for sectors 56, 73, 82, 97, 122, 131 and 148 in 1985 and for sectors 70 and 162 in the year 2000, but the differences are slight and none of the sectors affected takes a large share of PCE.

There are also, particularly in 1985, several sectors for which scenario E shows either the highest or the lowest spending per person. These are sectors in which the composition effect outweighs the income effect on one side of scenario E but reinforces it on the other side. In 1985, the number of primary units is identical in all three scenarios, as is the number of childless families and the number of families with children. The scenarios differ only in the number of children per family and thus the average household size (as table 2-4 indicates). The sectors for which expenditure per capita varies non-monotonically with income per capita are ones for which spending is quite sensitive to the number of children: 39 and 70 (clothing and shoes), 43 and 45 (lumber and furniture), 133 and 139 (vehicles and cycles), 151, 154, 155 and 169 (public transportation and lodging) and several heterogeneous sectors including 28, 35, 38, 54, 70, 80, 87, 128, 146, 147 and 174.

Differences in per capita spending between scenario E and either D or F rise slightly through time as income differences widen, but remain quite small for most sectors. In 1985, only nine sectors show differences, in either direction, in excess of 10 percent. By the year 2025, 27 sectors show differences of 10 percent or more, but there are still several sectors in which there is essentially no difference among scenarios.

Relatively little can be learned from analysis across the different population assumptions, but that only reflects the fact that the difference between two growth paths cannot be neatly partitioned at a point in time. The difference at some future date is a cumulative effect of different evolutions along alternative paths. This corresponds to the kind of choice society makes between one path and another, since it is not possible in one year to jump from one to another of two paths which have already diverged considerably. Instead, choices made at one time can only be evaluated by tracing their implications over alternative futures.

Summary of Demographic Effects

For each sector, we have calculated seven parameters for each year or period and each scenario. Five of these—the three components of growth due to scale, composition and income, and the two elasticities with respect to population and income—refer to change through time and involve both the growth and the no-growth projections. The other two—the cross-section demographic elasticity and the percentage difference in per capita expenditure—refer to differences between scenario E and scenarios D and F, and are based on the growth projections only; they are, by definition, not calculated for scenario E. We have summarized the results of these calculations and their implications for the effects of demographic growth and change, in the discussion accompanying tables 6-10 through 6-19. Now we present, in table 6-20, the mean and standard deviation of each parameter across sectors. (The number of sectors involved varies slightly, as some sectors disappear before the end of the projection period.) These figures show how the different parameters vary among sectors: note that

Table 6-20. Means and Standard Deviations of Decomposition and Elasticity Parameters, Across Sectors, by Scenario and Year or
Period

(Terminal) Year		Scenario D Mean	Std Dev	E Mean	Std Dev	F Mean	Std Dev
1985	SCA	38.0	18.7	32.6	22.1	25.9	19.1
through	COM	26.4	25.8	36.2	30.7	39.8	31.3
time	INC	35.6	27.7	31.2	32.8	34.3	34.9
	e_Y	0.618	1.030	0.525	0.977	0.488	0.828
	e_N	1.790	1.381	2.275	1.946	2.799	2.583
across	$e_N(E)$	0.02	2.42	*******	*******	−0.87	3.13
scenarios	%Diff	−3.02	6.31	*******	*******	4.15	7.56
	$/person						
2000	SCA	31.2	18.5	24.5	17.5	17.5	12.7
through	COM	14.2	16.7	14.6	15.2	18.3	17.2
time	INC	54.7	21.1	60.9	22.8	64.2	22.3
	e_Y	0.713	0.592	0.736	0.566	0.726	0.540
	e_N	1.500	0.801	1.650	0.984	2.053	1.354
across	$e_N(E)$	0.21	0.58	*******	*******	0.06	0.57
scenarios	%Diff	−5.87	4.29	*******	*******	5.21	3.24
	$/person						
2025	SCA	28.9	19.7	22.8	18.6	14.0	12.1
through	COM	4.2	12.7	7.5	11.4	11.3	15.2
time	INC	66.9	15.8	69.7	22.1	74.6	22.1
	e_Y	0.828	0.513	0.805	0.529	0.809	0.516
	e_N	1.221	0.467	1.347	0.593	1.694	1.019
across	$e_N(E)$	0.72	0.45	*******	*******	0.52	0.37
scenarios	%Diff	−4.88	7.46	*******	*******	6.88	5.70
	$/person						

they are unweighted calculations which treat all sectors equally rather
than in proportion to expenditure. In contrast, the same statistics calculated for large categories, in chapter 1, can be thought of as weighted
averages over the component sectors.

The results presented in table 6-20 show a number of tendencies, some
of which were noted earlier. First, the income component of growth becomes
more important through time, while the population scale component shrinks
and the effect of changing population composition shrinks still more,

nearly disappearing in scenarios D and E. Second, the elasticities of
growth through time converge toward unity, the income elasticity being
initially much lower (around 0.5 or 0.6) and the population elasticity
much higher (between 1.8 and 2.8). Third, the differences across sce-
narios--indicated by the cross-section elasticity or by the difference in
per capita spending--tend to increase through time, but never become very
large.

Fourth, as is shown by the standard deviations of the parameters,
variation among sectors tends to diminish through time. This tendency is
less marked than some of those indicated by the means of the parameters,
but it is still notable for most of the calculations, and is especially
clear in the case of the three elasticities e_Y, e_N and $e_N(E)$. Regardless
of whether the mean is increasing or decreasing, the variation in each of
these parameters tends toward zero. A similar but less pronounced effect
may be seen in the income share of growth, where the mean always rises
but the standard deviation always falls. For the composition effect, in
contrast, the standard deviation also declines, but less rapidly than the
mean: relative variation increases over time, as the expenditure sectors
become separated into those with and those without strong composition effects on growth. Fifth, the standard deviation is usually of the same
order of magnitude as the mean, at least initially.

This completes our review of the sector-by-sector analysis of income
and demographic effects of possible future growth. The reader who wishes
to study different aggregates of expenditure--total goods, for example,
or total consumer durables--can use the sectoral projections in tables
6-2 through 6-7 and the parameters defined in chapter 1 to repeat, modify
or extend our analysis. The results can also be used, in conjunction with
an input-output table or other exogenous models, to derive projections of
investment, analyze price effects, or otherwise carry the analysis beyond
the structure of personal consumption expenditure.

Finally, we must remind the reader that the actual course of future
income and population growth may not closely resemble any of the projected
scenarios we have used; and that even if income and population change according to one of these scenarios, changes in prices, tastes and technology
could lead to a structure of demand different from what we project. The

conclusions we have drawn seem nonetheless to be reasonably robust, bar-
ring any changes which are both very large and affect many sectors of the
economy. To reiterate the chief conclusion drawn from the analysis in
chapter 1, population size obviously affects the level of spending on almost every sector of demand, but even very large differences in population
growth and eventual population size do not seem likely to change the structure of demand dramatically. Major effects of population composition are
concentrated in relatively few sectors, and over time, most population
effects are outweighed by the changes that accompany the steady growth of
income.

248

References

ACIR American Council on Inter-governmental Relations. 1977. _Significant Features of Fiscal Federalism_, 1976-77 Edition (Washington, D.C., ACIR).

Almon, Clopper, Jr., Margaret B. Buckler, Lawrence M. Horwitz, and Thomas C. Reimbold. 1974. _1985: Interindustry Forecasts of the American Economy_ (Lexington, Mass., D.C. Heath).

Alterman, Jack, n.d. "Projections of Labor Force, Labor Productivity, Gross National Product and Households: Methodology and Data," (Washington, D.C.: Resources for the Future, Working Paper).

Anderson, Ronald, Joanna Kravits, and Odin W. Anderson. 1975. _Equity in Health Services_ (Cambridge, Mass., Ballinger).

Anderson, Ronald, Joanna Lion and Odin W. Anderson. 1976. _Two Decades of Health Services_ (Cambridge, Mass., Ballinger).

Appleman, Jack. 1972. "Health Care," in _Economic Aspects of Population Change, Report of the Commission on Population Growth and the American Future, Volume 2_ (Washington, D.C., GPO).

Bureau of the Census. Various Years. "School Enrollment," Series P-20, _Current Population Reports_ (Washington, D.C., GPO).

Bureau of the Census. 1972. "Projections of School and College Enrollment: 1971 to 2000," _Current Population Reports_, Series P-25, No. 473 (Washington, D.C., GPO).

Bureau of the Census. 1975. "Value of New Construction Put in Place, 1947 to 1974," _Construction Reports_, Series C30, Number 745 (Washington, D.C., GPO)

Butz, William P. and Paul L. Jordan. 1972. "Education" in _Economic Aspects of Population Change, Report of the Commission on Population Growth and the American Future, Volume 2_ (Washington, D.C., GPO).

Carnegie Commission on Higher Education. 1972. _The More Effective Use of Resources_ (McGraw-Hill).

Espenshade, Thomas J. 1977. "How a Trend Toward a Stationary Population Affects Consumer Demand," Seventeenth General Conference of the International Union for the Scientific Study of Population (Mexico City: August).

Freeland, Mark S., Barry J. Greengart, and Myron J. Katzoff. 1977. "Projections of National Health Expenditures 1977-82." Unpublished note, National Health Insurance Modeling Group, Dept. of Health, Education and Welfare.

Fuchs, Victor R. and Marcia J. Kramer. 1972. _Determinants of Expenditures for Physicians' Services in the United States 1948-68_ (Washington, Department of Health, Education and Welfare Publication No. (HSM) 73-3013).

249

Gibson, Robert M. and Marjorie Smith Mueller. 1977. "National Health Expenditures, Fiscal Year 1976," Social Security Bulletin, April, 1977, pp. 3-22.

Gibson, Robert M., Marjorie Smith Mueller, and Charles R. Fisher. 1977. "Age Differences in Health Care Spending, Fiscal Year 1976," Social Security Bulletin, August, 1977, pp. 3-14.

Herzog, Henry, n.d., "A Method for Recognizing 'Demographic-Specific' Family Expenditure Pattern Shifts in Long-Run Personal Consumption Forecasting" (Washington, D.C.: Resources for the Future).

Houthakker, Hendrick S. and Lester D. Taylor. 1970. Consumer Demand in the United States: Analyses and Projections (Cambridge: Harvard University Press).

International Research and Technology Corporation. 1976. "Automobile Forecasting Models," report prepared for the Office of Technology Assessment, U.S. Congress by R.U. Ayers, et. al. (September).

Kasper, Judith. 1975. "Physician Utilization and Family Size," in Ronald Andersen, Joanna Kravits and Odin W. Anderson, Equity in Health Services (Ballinger: Cambridge, Mass., 1975).

Lluch, Constantino, Alan Powell and Ross Williams. 1977. Patterns in Household Demand and Saving (New York: Oxford University Press for the World Bank).

Marcin, Thomas C. 1974. The Effects of Declining Population Growth on the Demand for Housing, USDA Forest Service General Technical Report NC-11 (Washington, GPO).

Mathematica Policy Research, Inc. 1977. Unpublished projections of population and household structure (Washington, D.C.).

Musgrove, Philip. 1979. "Permanent Household Income and Consumption in Urban South America," American Economic Review 69 (June).

National Center for Education Statistics, Various Years, a. Biennial Survey of Education in the United States (Washington, GPO).

National Center for Education Statistics, Various Years, b. Digest of Education Statistics (Washington, GPO).

National Center for Education Statistics, Various Years, c. Statistics of State School Systems (Washington, GPO).

National Center for Education Statistics. 1977. Projections of Education Statistics to 1985-86 (Washington, GPO).

Paglin, Morton. 1975. "The Measurement and Trend of Inequality: A Basic Revision," American Economic Review 65 (September).

Reid, Margaret G. 1962. Housing and Income (Chicago, University of Chicago Press).

Ridker, Ronald G. and William D. Watson. 1980. To Choose a Future: Resource and Environmental Consequences of Alternative Growth Paths (Washington, D.C.: Resources for the Future).

Serow, William J. and Thomas J. Espenshade. 1977. "The Economics of De-
    clining Population Growth: An Assessment of the Current Literature,"
    Conference on the Economic Consequences of Slowing Population Growth,
    Center for Population Research, National Institute of Child Health
    and Human Development (Washington, D.C., May).

Shapanka, Adele. 1978. "Long-Range Technological Forecasts for Use in
    Studying the Resource and Environmental Consequences of U.S. Popula-
    tion and Economic Growth, 1975-2025," Resources for the Future Dis-
    cussion Paper D-31 (May).

Smith, Ron. 1976. Consumer Demand for Cars in the U.S.A. (Cambridge:
    Cambridge University Press).

Social Security Administration. 1976. Research and Statistics Note (note
    No. 5) (Washington, GPO).

Stoikov, Vladimir. 1975. "How Misleading are Income Distributions?",
    Review of Income and Wealth 21 (June).

Taylor, Lester D. 1971. "The Personal Tax Surcharge and Consumer Demand,
    1968-70," Brookings Papers on Economic Activity 1.

Trapnell, Gordon R. 1976. "A Comparison of the Costs of Major National
    Health Insurance Proposals" (Washington, National Technical Informa-
    tion Service).

U.S. Census Bureau, Current Population Reports. Various Years. Series
    P-25 No. 493 (December 1972), No. 601 (October 1975), No. 607 (August,
    1975) and No. 704 (July, 1977).

About the author

Philip Musgrove has a long association with both Resources for the Future and the Brookings Institution, where he has worked on consumption and income studies for the United States and Latin America. He has taught courses in economics and statistics at American University, George Washington University, and the University of Florida.

He has also been a consultant for the World Bank, the U.S. Department of Agriculture, and the central banks of Venezuela and the Dominican Republic. His published papers deal with consumer behavior, poverty, and income distribution. His previous RFF publication, written with Joseph Grunwald, is <u>Natural Resources in Latin American Development</u>. This work illustrates regional and national issues, explores the nature of government participation and control, and examines the influence of foreign private capital.

For Product Safety Concerns and Information please contact our EU
representative GPSR@taylorandfrancis.com
Taylor & Francis Verlag GmbH, Kaufingerstraße 24, 80331 München, Germany